OPTIMAL SPACECRAFT TRAJECTORIES

Optimal Spacecraft Trajectories

JOHN E. PRUSSING

Professor Emeritus
Department of Aerospace Engineering
University of Illinois at Urbana-Champaign
Urbana, Illinois, USA

OXFORD
UNIVERSITY PRESS

OXFORD
UNIVERSITY PRESS

Great Clarendon Street, Oxford, OX2 6DP,
United Kingdom

Oxford University Press is a department of the University of Oxford.
It furthers the University's objective of excellence in research, scholarship,
and education by publishing worldwide. Oxford is a registered trade mark of
Oxford University Press in the UK and in certain other countries

First Edition published in 2018

Impression: 1

Published in the United States of America by Oxford University Press
198 Madison Avenue, New York, NY 10016, United States of America

British Library Cataloguing in Publication Data
Data available

Library of Congress Control Number: 2017959440

ISBN 978–0–19–881108–4 (hbk.)
ISBN 978–0–19–881111–4 (pbk.)

Printed and bound by
CPI Group (UK) Ltd, Croydon, CR0 4YY

Preface

This text takes its title from an elective course at the University of Illinois at Urbana-Champaign that has been taught to graduate students for the past 29 years. The book assumes the reader has a background of undergraduate engineering (aerospace, mechanical, or related), physics, or applied mathematics. Parts of the book rely heavily on Optimal Control Theory, but an entire chapter on that topic is provided for those unfamiliar with the subject.

Books on the field of optimal spacecraft trajectories are very few in number and date back to the 1960s, and it has been nearly 40 years since a comprehensive new book has appeared. Hence this volume.

Some classical results are presented using modern formulations and both impulsive-thrust and continuous-thrust trajectories are treated. Also included are topics not normally covered, such as cooperative rendezvous and second-order conditions. An unusually large number of appendices (7) is provided to supplement the main text.

The book is suitable for a one-semester graduate-level university course. It is also a scholarly reference book. Problems are included at the ends of the chapters and some of the appendices. They either illustrate the subject matter covered or extend it. Solving them will enhance the reader's understanding of the material, especially if they are assigned by your instructor as homework!

I am indebted to all the students who have taken my graduate course and to the many colleagues I have interacted with over the years.

I have greatly benefitted from interacting with some of the giants in optimal trajectories: Derek F. Lawden, Theodore N. Edelbaum, John V. Breakwell, and Richard H. Battin. These fine gentlemen have passed on, but I hope this book helps convey their contributions, both directly and indirectly.

J.E.P.
Urbana. Illinois, USA
February, 2017

Contents

Introduction

Performance index for an optimal trajectory

The performance index for an optimal spacecraft trajectory is typically minimum propellant consumption, not because of the monetary cost of the propellant, but because for each kilogram of propellant saved an additional kilogram of payload is delivered to the final destination. The term "minimum fuel" is often used in place of "minimum propellant", even for chemical rockets in which the propellant is composed of both fuel and oxidizer. As is often the case, the term "fuel" is used rather than "propellant" because it's shorter – having one syllable instead of three. In some applications other performance indices are used, such as minimum time, maximum range, etc.

In optimizing spacecraft trajectories the two major branches of optimization theory are used: parameter optimization and optimal control theory. In parameter optimization the parameters are constants and we minimize a function of a finite number of parameters. An example is to represent the propellant consumed as the sum of a finite number of velocity changes caused by rocket thrusts. By contrast, optimal control is a problem of infinite dimensions where the variables are functions of time. An example of this is the fuel consumed by continuously varying the thrust magnitude for a finite duration rocket thrust.

Types of orbital maneuvers

The three types of orbital maneuvers are orbital interception, orbital rendezvous, and orbit transfer. In orbital interception at the final time t_f the position vector of the spacecraft $\mathbf{r}(t_f)$ and the position vector of a target body $\mathbf{r}^{\#}(t_f)$ are equal: $\mathbf{r}(t_f) = \mathbf{r}^{\#}(t_f)$. A word of explanation of the term "target" would be helpful here. The term is more general (and more benign) than describing a military target to be eliminated. For a planetary flyby the target body is the flyby planet, such as Jupiter during the Cassini Mission to Saturn. At the flyby time the spacecraft position vector and the flyby planet position vector are equal and the spacecraft has intercepted the flyby planet.

Optimal Spacecraft Trajectories. John E. Prussing.

By contrast, in an orbital rendezvous both the position vector and the velocity vector of the spacecraft at the final time are equal to those of the target body: $\mathbf{r}(t_f) = \mathbf{r}^{\#}(t_f)$ and $\mathbf{v}(t_f) = \mathbf{v}^{\#}(t_f)$. Putting a lander on Mars is an example of a rendezvous of a spacecraft with a planet.

Lastly, in an orbit transfer there is no target body. The desired final condition is a specific orbit, characterized by values of semimajor axis, eccentricity, inclination, etc.

Any one of these maneuvers can be either time-fixed (constrained, closed, finite or fixed horizon) or time-open (unconstrained, open, infinite horizon). Note that in a well-posed time-open optimization problem the value of t_f will be optimized as part of the solution.

Notation and preliminaries

A bold-faced symbol denotes a vector (column matrix) $\mathbf{x}$ having components (elements) x_i. An example is the n-vector

$$\mathbf{x} = \begin{bmatrix} x_1 \\ x_2 \\ \cdot \\ \cdot \\ x_n \end{bmatrix} \tag{1}$$

A symbol x without the bold face is the magnitude of the vector $\mathbf{x}$:

$$x = |\mathbf{x}| = (\mathbf{x} \cdot \mathbf{x})^{\frac{1}{2}} = (\mathbf{x}^T\mathbf{x})^{\frac{1}{2}} \tag{2}$$

We will encounter a scalar function of a vector $\phi(\mathbf{x})$, e.g., a gravitational potential function, and our convention is that the gradient $\frac{\partial\phi}{\partial\mathbf{x}}$ is a row $(1 \times n)$ vector:

$$\frac{\partial\phi}{\partial\mathbf{x}} = \left[\frac{\partial\phi}{\partial x_1}\,\frac{\partial\phi}{\partial x_2}\cdots\frac{\partial\phi}{\partial x_n}\right] \tag{3}$$

One advantage of this convention is that

$$d\phi = \frac{\partial\phi}{\partial\mathbf{x}}d\mathbf{x} \tag{4}$$

with no transpose symbol being needed. The gradient appearing in Eq. (4) corresponds to the gradient vector $\mathbf{g} = \left(\frac{\partial\phi}{\partial\mathbf{x}}\right)^T$, which is the vector normal to the surface ϕ = constant.

We will also encounter a *Jacobian matrix,* which is the derivative of a vector $\mathbf{F}$ with respect to another vector $\mathbf{x}$. If $\mathbf{F}$ is a p-vector and $\mathbf{x}$ is an n-vector the Jacobian is an $p \times n$ matrix with ij element equal to $\frac{\partial F_i}{\partial x_j}$, where i is the row index and j is the column index:

$$\frac{\partial \mathbf{F}}{\partial \mathbf{x}} = \begin{bmatrix} \frac{\partial F_1}{\partial x_1} & \frac{\partial F_1}{\partial x_2} & \cdot & \frac{\partial F_1}{\partial x_n} \\ \frac{\partial F_2}{\partial x_1} & \frac{\partial F_2}{\partial x_2} & \cdot & \frac{\partial F_2}{\partial x_n} \\ \cdot & \cdot & \cdot & \cdot \\ \frac{\partial F_p}{\partial x_1} & \frac{\partial F_p}{\partial x_2} & \cdot & \frac{\partial F_p}{\partial x_n} \end{bmatrix} \tag{5}$$

Finally, we distinguish between a function and a functional. A function maps numbers into numbers. An example is $f(x) = x^2$ which for $f(2) = 4$. By contrast a functional maps functions into numbers. An example is $J = \int_0^1 f(x)dx$, which for $f(x) = x^2$ yields $J = 1/3$.

1 Parameter Optimization

1.1 Unconstrained parameter optimization

Parameter optimization utilizes the theory of ordinary maxima and minima. In our analysis, we will use the notation in Chapter 1 of Ref. [1.1]. The unconstrained problem is to determine the value of the m-vector $\mathbf{u}$ of independent parameters (decision variables) to minimize the function

$$L(\mathbf{u}) \tag{1.1}$$

where the scalar L is termed a performance index or cost function. We note that a maximum value of L can be determined by minimizing $-L$, so we will treat only minimization conditions.

If the u_i are independent (no constraints) and the first and second partial derivatives of L are continuous, then a stationary solution $\mathbf{u}^*$ satisfies the *necessary conditions* (NC) that

$$\frac{\partial L}{\partial \mathbf{u}} = \mathbf{0}^T \tag{1.2a}$$

where $\mathbf{0}$ is the zero vector (every element equal to zero) and that the *Hessian matrix*

$$\left(\frac{\partial^2 L}{\partial \mathbf{u}^2}\right)_{\mathbf{u}^*} \geq 0 \tag{1.2b}$$

which is a shorthand notation that the $m \times m$ matrix having elements $\frac{\partial^2 L}{\partial u_i \partial u_j}$ evaluated at $\mathbf{u}^*$ must be positive semidefinite (all eigenvalues zero or positive). Equation (1.2a) represents m equations that determine the values of the m variables u_i^*.

Sufficient conditions (SC) for a local minimum are the stationary condition (1.2b) and that the Hessian matrix of (1.2b) be positive definite (all eigenvalues positive). (The

Optimal Spacecraft Trajectories. John E. Prussing.

semidefinite NC is simply a statement that for the stationary point to be a minimum, it is necessary that it is not a maximum!)

In the general case of m parameters we can write the first variation of our function as

$$\delta L = \frac{\partial L}{\partial u_1}\delta u_1 + \cdots + \frac{\partial L}{\partial u_m}\delta u_m = 0 \tag{1.3}$$

and conclude, using the conditions of Eq. (1.2a), that at a stationary point $\delta L = 0$. In a series expansion about the stationary point:

$$L(\mathbf{u}) = L(\mathbf{u}^*) + \delta L + \delta^2 L + \cdots \tag{1.4}$$

the term $\delta^2 L$ is the second variation. A zero value for the first variation and a positive value for the second variation are the SC for a local minimum.

As an example consider $\mathbf{u}$ to be a 2-vector. The NC (1.2a) for a minimum provides two equations that determine the values of u_1^* and u_2^*.

The SC is

$$\delta^2 L = \delta\mathbf{u}^T \left(\frac{\partial^2 L}{\partial \mathbf{u}^2}\right)_{\mathbf{u}^*} \delta\mathbf{u} = [\delta u_1 \delta u_2] \begin{bmatrix} \dfrac{\partial^2 L}{\partial u_1 \partial u_1} & \dfrac{\partial^2 L}{\partial u_1 \partial u_2} \\ \dfrac{\partial^2 L}{\partial u_2 \partial u_1} & \dfrac{\partial^2 L}{\partial u_2 \partial u_2} \end{bmatrix}_{\mathbf{u}^*} \begin{bmatrix} \delta u_1 \\ \delta u_2 \end{bmatrix} > 0 \tag{1.5}$$

where $\delta u_1 \equiv u_1 - u_1^*$ and $\delta u_2 \equiv u_2 - u_2^*$ are arbitrary infinitesimal variations away from their stationary values. If the strict inequality in Eq. (1.5) is satisfied for all nonzero δu_1 and δu_2 the matrix is positive definite. A simple test is that a matrix is positive definite if all its leading principal minors are positive. For the 2×2 matrix in Eq. (1.5) the required conditions are that the first principal minor is positive:

$$\frac{\partial^2 L}{\partial u_1 \partial u_1} > 0 \tag{1.6a}$$

and that the second principal minor (the determinant of the matrix in this case) is positive:

$$\frac{\partial^2 L}{\partial u_1 \partial u_1}\frac{\partial^2 L}{\partial u_2 \partial u_2} - \left(\frac{\partial^2 L}{\partial u_1 \partial u_2}\right)^2 > 0 \tag{1.6b}$$

As a very simple example consider

$$L(\mathbf{u}) = 2u_1^2 + u_2^2 + u_1 u_2 \tag{1.7}$$

The reader can verify that the NC (1.2a) for this example has the unique solution $u_1 = u_2 = 0$ and that Eqs (1.6a) and (1.6b) are also satisfied because the values of the leading principal minors are 4 and 7. So the function has a minimum at $\mathbf{u} = \mathbf{0}$.

1.2 Parameter optimization with equality constraints

If there are one or more constraints the values of the parameters are not independent, as we assumed previously. As an example, if a point (x_1, x_2) is constrained to lie on a circle of radius R centered at the origin we have the constraint

$$x_1^2 + x_2^2 - R^2 = 0 \tag{1.8}$$

The constrained minimum problem is to determine the values of *m-decision parameters* $\mathbf{u}$ that minimize the scalar cost $L(\mathbf{x}, \mathbf{u})$ of $n + m$ parameters, where the *n-state parameters* $\mathbf{x}$ are determined by the decision parameters through a set of n equality constraints:

$$\mathbf{f}(\mathbf{x}, \mathbf{u}) = \mathbf{0} \tag{1.9}$$

where $\mathbf{f}$ is an n-vector. Note that the number of state parameters n is *by definition* equal to the number of constraints. In general, $n > m$ so there are $n - m$ independent variables. (If $m = n$ the problem is overconstrained and there is no optimization possible; there is either only one solution or no solution.)

1.2.1 Lagrange multipliers

In the constrained problem the parameters can be treated as independent if we introduce n additional variables $\lambda_1, \lambda_2, \ldots, \lambda_n$, called *Lagrange multipliers*. We use these Lagrange multipliers to adjoin the constraints to the cost function by forming an augmented function H

$$H(\mathbf{x}, \mathbf{u}, \boldsymbol{\lambda}) = L(\mathbf{x}, \mathbf{u}) + \boldsymbol{\lambda}^T \mathbf{f}(\mathbf{x}, \mathbf{u}) \tag{1.10}$$

where $\boldsymbol{\lambda}^T \equiv [\lambda_1\ \lambda_2 \cdots \lambda_n]$. From Eq. (1.10) we see that if the constraints are satisfied, $\mathbf{f} = \mathbf{0}$, H is equal to L, and for first-order NC we can treat the problem as the unconstrained optimization $\delta H = 0$. The necessary conditions are

$$\frac{\partial H}{\partial \mathbf{x}} = \mathbf{0}^T \tag{1.11a}$$

$$\frac{\partial H}{\partial \mathbf{u}} = \mathbf{0}^T \tag{1.11b}$$

and the constraint

$$\frac{\partial H}{\partial \boldsymbol{\lambda}} = \mathbf{f}^T = \mathbf{0}^T \tag{1.11c}$$

Equations (1.11a–c) are $2n + m$ equations in $2n + m$ unknowns ($\mathbf{x}$, $\boldsymbol{\lambda}$, and $\mathbf{u}$).

Introducing the n additional variables $\boldsymbol{\lambda}$ seems to be counterproductive, because we now have more unknowns to solve for. Instead of introducing the Lagrange multipliers we could, in principle, solve the constraints $\mathbf{f} = \mathbf{0}$ for the n state variables $\mathbf{x}$ in terms of the decision variables $\mathbf{u}$, leaving only m equations to solve for the m decision variables. However, those equations, while fewer, are usually more complicated than Eqs (1.11a–c), so it is easier to solve the simpler $2n + m$ equations (see Problem 1.2).

The Lagrange multiplier $\boldsymbol{\lambda}$ has an interesting and useful interpretation, which makes it worth solving for. It provides the sensitivity of the stationary cost to small changes in the constants in the constraint equations. So, we can obtain the change in the stationary cost due to small changes in these constants without resolving the problem. This can be very useful in a complicated problem that is solved numerically.

This interpretation is discussed in Appendix A. To summarize, write the constraints as

$$\mathbf{f}(\mathbf{x}, \mathbf{u}, \mathbf{c}) = \mathbf{0} \tag{1.12}$$

where $\mathbf{c}$ is a q-dimensional vector of constants in the constraint equations. Due to changes in these constants, $d\mathbf{c}$, the change in the stationary cost is given by

$$dL^* = \boldsymbol{\lambda}^T \mathbf{f}_c d\mathbf{c} \tag{1.13}$$

where $\mathbf{f}_c = \frac{\partial \mathbf{f}}{\partial \mathbf{c}}$ is an $n \times q$ matrix.

The SC analogous to the Hessian matrix condition $\left(\frac{\partial^2 L}{\partial \mathbf{u}^2}\right)_{\mathbf{u}^*} > 0$ for the unconstrained problem is given in Section 1.3 of Ref. [1.1] as

$$\left(\frac{\partial^2 L}{\partial \mathbf{u}^2}\right)_{\mathbf{f}=0} = H_{\mathbf{uu}} - H_{\mathbf{ux}}\mathbf{f}_{\mathbf{x}}^{-1}\mathbf{f}_{\mathbf{u}} - \mathbf{f}_{\mathbf{u}}^T\mathbf{f}_{\mathbf{x}}^{-T}H_{\mathbf{xu}} + \mathbf{f}_{\mathbf{u}}^T\mathbf{f}_{\mathbf{x}}^{-T}H_{\mathbf{xx}}\mathbf{f}_{\mathbf{x}}^{-1}\mathbf{f}_{\mathbf{u}} > 0 \tag{1.14}$$

where the symbol $\mathbf{f}_x^{-T}$ is shorthand for the transpose of the inverse, which is equal to the inverse of the transpose (see Problem 1.3).

Example 1.1 Minimization using a Lagrange multiplier

Determine the rectangle of maximum perimeter that can be inscribed in a circle of radius R. In this example, there are two variables x and y and a single constraint—the equation of the circle. So $n = m = 1$ and we arbitrarily assign y to be the decision variable u. Let the center of the circle lie at $(x, u) = (0, 0)$ and denote the corner of the rectangle in the first quadrant as (x, u). To maximize the perimeter, we minimize its negative and form

$$L(x, u) = -4(x + u) \tag{1.15}$$

subject to the constraint

$$f(x, u) = x^2 + u^2 - R^2 = 0 \tag{1.16}$$

where x and u are scalars.

We construct the augmented function H as

$$\begin{aligned} H(x,u,\lambda) &= L(x,u) + \lambda f(x,u) \\ &= -4(x+u) + \lambda(x^2 + u^2 - R^2) \end{aligned} \tag{1.17}$$

Applying the NC of Eqs. (1.11a–c):

$$H_x = \frac{\partial H}{\partial x} = -4 + 2\lambda x = 0 \tag{1.18a}$$

$$H_u = \frac{\partial H}{\partial u} = -4 + 2\lambda u = 0 \tag{1.18b}$$

and the constraint

$$H_\lambda = \frac{\partial H}{\partial \lambda} = x^2 + u^2 - R^2 = 0 \tag{1.18c}$$

where we use subscript notation for partial derivatives. Equations (1.18a–c) are easily solved to yield

$$x^* = u^* = \frac{\sqrt{2}}{2}R \tag{1.19}$$

$$\lambda = \frac{2\sqrt{2}}{R} \tag{1.20}$$

where Eq. (1.19) indicates that the rectangle of maximum perimeter is a square. To verify that our stationary solution is indeed a minimum we use Eq. (1.14):

$$\begin{aligned} \left(\frac{\partial^2 L}{\partial u^2}\right)_{f=0} &= H_{uu} - H_{ux} f_x^{-1} f_u - f_u f_x^{-1} H_{xu} + f_u f_x^{-1} H_{xx} f_x^{-1} f_u \\ &= \frac{4\sqrt{2}}{R} - 0 - 0 + \frac{4\sqrt{2}}{R} = \frac{8\sqrt{2}}{R} > 0 \end{aligned} \tag{1.21}$$

and this constrained second derivative being positive satisfies the SC for a minimum, resulting in a maximum perimeter of $4\sqrt{2}R = 5.657R$.

Applying Eq. (1.13) with the single constant c being R:

$$dL^* = \lambda f_R dR = \frac{2\sqrt{2}}{R}(-2R)dR = -4\sqrt{2}dR \tag{1.22}$$

Consider a nominal value $R = 1$ and a change in the radius of $dR = 0.1$. Using Eq. (1.22) the change in the minimum cost is equal to $-0.4\sqrt{2} = -0.566$, resulting in a maximum perimeter of $5.657 + 0.566 = 6.223$. This compares very favorably with the exact value of $4\sqrt{2}(1.1) = 6.223$

For this simple example using the interpretation of the Lagrange multiplier is not very profound, but in a complicated problem, especially if the solution is obtained numerically rather than analytically, a good approximation to the cost of a neighboring stationary solution can be obtained without completely resolving the problem.

Appendix B provides a more complicated constrained minimum example: the Hohmann transfer as a minimum Δv solution.

1.3 Parameter optimization with an inequality constraint

Consider the simple case of a scalar cost $L(\mathbf{y})$ subject to a constraint of the form $f(\mathbf{y}) \leq 0$, where $\mathbf{y}$ is p dimensional and f is a scalar. We will consider only the simple case of a single constraint, because that is all we need for the single application that we treat in Section 7.2. Consideration of the general case results in the Karush–Kuhn–Tucker conditions and is discussed in Section 1.7 of Ref. [1.1] (as the Kuhn–Tucker conditions).

As in the equality constraint case in Section 1.1 a Lagrange multiplier λ is introduced and a scalar H is defined as:

$$H(\mathbf{y}, \lambda) = L(\mathbf{y}) + \lambda f(\mathbf{y}) \tag{1.23}$$

Note that there is a total of $p + 1$ unknowns, namely the elements of $\mathbf{y}$ and λ. As in the equality constraint case an NC can be written as

$$\frac{\partial H}{\partial \mathbf{y}} = \mathbf{0}^T \tag{1.24}$$

but, unlike the equality constraint case, the inequality constraint is either *active* or *inactive*, resulting in the NC:

$$\lambda \geq 0 \quad \text{if } f(\mathbf{y}) = 0 \quad \text{(active)} \tag{1.25}$$

$$\lambda = 0 \quad \text{if } f(\mathbf{y}) < 0 \quad \text{(inactive)} \tag{1.26}$$

Note that in either case the product $\lambda f = 0$, resulting in H being equal to L.

Expanding Eq. (1.24):

$$\frac{\partial H}{\partial \mathbf{y}} = \frac{\partial L}{\partial \mathbf{y}} + \lambda \frac{\partial f}{\partial \mathbf{y}} = \mathbf{0}^T \tag{1.27}$$

or

$$\frac{\partial L}{\partial \mathbf{y}} = -\lambda \frac{\partial f}{\partial \mathbf{y}} \tag{1.28}$$

which provides the geometrical interpretation that when the constraint is active the gradients $\frac{\partial L}{\partial \mathbf{y}}$ and $\frac{\partial f}{\partial \mathbf{y}}$ are oppositely directed (see Fig. 1.1 for $p = 2$). From that figure we see

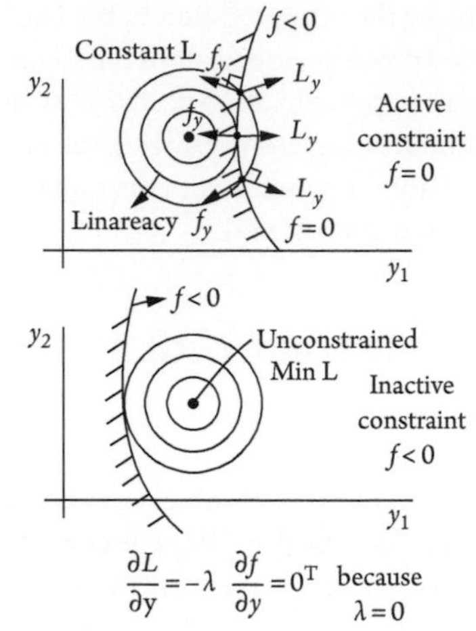

Figure 1.1 An active constraint and an inactive constraint

that at any point on the constraint not at the constrained minimum (where the gradients are oppositely directed) we can slide along the constraint and decrease the cost until we reach the constrained minimum.

This is even more evident when we note that the SC for the constraint being active is $\lambda > 0$, indicating that the only way to decrease the cost is to violate the constraint.

If the constraint is inactive the constrained minimum is the unconstrained minimum. In that case $\frac{\partial L}{\partial \mathbf{y}} = \mathbf{0}^T$ as seen in Eq. (1.28) with $\lambda = 0$.

Example 1.2 Simple inequality constraint

In order to show how the NC are applied, consider the very simple example to minimize

$$L(y) = \frac{1}{2}y^2 \tag{1.29}$$

subject to the constraint

$$f(y) = y - a \leq 0 \tag{1.30}$$

where y and a are scalars. Use Eq. (1.23) to form H:

$$H(y, \lambda) = \frac{1}{2}y^2 + \lambda(y - a) \tag{1.31}$$

Equation (1.24) provides the NC:

$$\frac{\partial H}{\partial y} = y + \lambda = 0 \tag{1.32}$$

where from Eqs. (1.25) and (1.26)

$$\lambda \geq 0 \quad \text{if} \quad y = a \quad (\text{active}) \tag{1.33}$$

$$\lambda = 0 \quad \text{if} \quad y < a \quad (\text{inactive}) \tag{1.34}$$

We next assume the constraint is active or inactive and investigate whether the NC are satisfied. Consider two separate cases, $a = 1$ and $a = -1$.

Case I $a = 1$

Assume the constraint is active. Then $y = 1$ and, from Eq. (1.32), $\lambda = -1$, which violates NC (1.33). Assuming the constraint is inactive yields $\lambda = 0$ and $y = 0$, which satisfy the NC. So, the constrained minimum is at $y = 0$ and is the same as the unconstrained minimum with a cost $L = 0$. Because the constraint is inactive the SC that $\partial^2 L/\partial y^2 = 1 > 0$ is satisfied.

Case II $a = -1$

Assume the constraint is inactive. Then $\lambda = 0$ and $y = 0$, which violates the constraint $y < -1$. Assuming the constraint is active yields $y = -1$ and $\lambda = 1$, which satisfy the NC. So the constrained minimum occurs at $y = -1$ with a cost $L = 1/2$. Note also that the SC $\lambda > 0$ is satisfied.

Section 7.2 provides a more complicated example of an inequality constraint minimization, namely, the terminal maneuver for an optimal cooperative impulsive rendezvous.

Problems

1.1 a) Determine the value of the *constant* parameter k that "best" approximates a given function $f(x)$ on the interval $a \leq x \leq b$. Use as a cost function to be minimized

$$J(k) = \frac{1}{2}\int_a^b [k - f(x)]^2 dx$$

This is the *integral-squared error*, which is the continuous version of the familiar *least-squares* approximation in the finite-dimensional case.

Verify that a global minimum is achieved for your solution.

b) For $a = 0$, $b = 1$, and $f(x) = x^3$ evaluate the cost J for (i) the optimal value of k and (ii) another value of k that you choose.

1.2 Solve the problem in Example 1.1 without using a Lagrange multiplier.

1.3 Show that for a nonsingular square matrix the inverse of its transpose is equal to the transpose of its inverse.

1.4* Minimize $L(\mathbf{y}) = \frac{1}{2}(y_1^2 + y_2^2)$ subject to the constraints $y_1 + y_2 = 1$ and $y_1 \geq 3/4$.

a) Attempt to satisfy the NC by assuming the inequality constraint is *inactive.*

b) Attempt to satisfy the NC by assuming the inequality constraint is *active.*

c) Determine whether the solution to the NC satisfies the SC.

Reference

[1.1] Bryson, A.E., Jr., and Ho, Y-C, *Applied Optimal Control.* Hemisphere Publishing, 1975.

2 Rocket Trajectories

2.1 Equations of motion

The equation of motion of a spacecraft which is thrusting in a gravitational field can be expressed in terms of the orbital radius vector **r** as:

$$\ddot{\mathbf{r}} = \mathbf{g}(\mathbf{r}) + \boldsymbol{\Gamma}; \qquad \boldsymbol{\Gamma} = \Gamma \mathbf{u} \tag{2.1}$$

The variable $\boldsymbol{\Gamma}$ is the thrust acceleration vector. The scalar Γ is the magnitude of the thrust acceleration defined as the thrust (force) T divided by the mass of the vehicle m. The variable **u** is a unit vector in the thrust direction, and $\mathbf{g}(\mathbf{r})$ is the gravitational acceleration vector. Equation (2.1) is somewhat deceptive and looks like a simple statement of "F equals ma" with the thrust term appearing on the right-hand side as if it were an external force like gravity. In actuality, F does **not** equal ma (but it does equal the time derivative of the linear momentum mv), because the mass is changing due to the generation of thrust and because the thrust is an *internal* force in the system defined as the combination of the vehicle and the exhausted particles. A careful derivation of Eq. (2.1) requires deriving the so-called *rocket equation* by equating the net external force (such as gravity) to the time rate of change of the linear momentum of the vehicle-exhaust particle system (see Sections 6.1–6.4 of Ref. [2.1]).

An additional equation expresses the change in mass of the spacecraft due to the generation of thrust:

$$\dot{m} = -b; \qquad b \geq 0 \tag{2.2}$$

In Eq. (2.2) b is the (nonnegative) mass flow rate. The thrust magnitude T is given by $T = bc$, where c is the *effective* exhaust velocity of the engine. The word "effective" applies in the case of high-thrust chemical engines where the exhaust gases may not be fully expanded at the nozzle exit. In this case an additional contribution to the thrust exists and the effective exhaust velocity is

$$c = c_a + (p_e - p_\infty)\frac{A_e}{b} \tag{2.3}$$

In Eq. (2.3) the subscript e refers to conditions at the nozzle exit, c_a is the actual (as opposed to effective) exhaust velocity, and p_∞ is the ambient pressure. If the gases are exhausted into the vacuum of space, $p_\infty = 0$.

An alternative to specifying the effective exhaust velocity is to describe the engine in terms of its *specific impulse*, defined to be:

$$I_{sp} = \frac{(bc)\Delta t}{(b\Delta t)g} = \frac{c}{g} \tag{2.4}$$

where g is the gravitational attraction at the earth surface, equal to 9.80665 m/s^2. The specific impulse is obtained by dividing the mechanical impulse delivered to the vehicle by the weight (on the surface of the earth) of propellant consumed. The mechanical impulse provided by the thrust force over a time Δt is simply $(bc)\Delta t$, and, in the absence of other forces acting on the vehicle is equal to the change in its linear momentum. The weight (measured on the surface of the earth) of propellant consumed during that same time interval is $(b\Delta t)g$, as shown in Eq. (2.4). Note that if instead one divided by the *mass* of the propellant (which is the fundamental measure of the amount of substance), the specific impulse would be identical to the exhaust velocity. However, the definition in Eq. (2.4) is in standard use with the value typically expressed in units of seconds.

2.2 High-thrust and low-thrust engines

A distinction between high- and low-thrust engines can be made based on the value of the nondimensional ratio $\Gamma_{\max}/g$. For high-thrust devices this ratio is greater than unity and thus these engines can be used to launch vehicles from the surface of the earth. This ratio extends as high as perhaps 100. The corresponding range of specific impulse values is between 200 and approximately 850 sec., with the lower values corresponding to chemical rockets, both solid and liquid, and the higher values corresponding to nuclear thermal rockets.

For low thrust devices the ratio $\Gamma_{\max}/g$ is quite small, ranging from approximately 10^{-2} down to 10^{-5}. These values are typical of electric rocket engines such as magnetohydrodynamic (MHD), plasma arc and ion devices, and solar sails. The electric engines typically require separate power generators such as a nuclear radioisotope generator or solar cells. The ratio for solar sails is of the order of 10^{-5}.

2.3 Constant-specific-impulse (CSI) and variable-specific-impulse (VSI) engines

Two basic types of engines exist: CSI and VSI, also called power-limited engines. The CSI category includes both high- and low- thrust devices. The mass flow rate b in some cases can be continuously varied but is limited by a maximum value $b_{\max}$. For this reason, this type of engine is also described as a thrust-limited engine, with $0 \leq \boldsymbol{\Gamma} \leq \boldsymbol{\Gamma}_{\max}$.

The VSI category includes those low thrust engines which need a separate power source to run the engine, such as an ion engine. For these engines, the power is limited by a maximum value P_{max}, but the specific impulse can be varied over a range of values. The fuel expenditure for the CSI and VSI categories is handled separately.

The equation of motion (1) can be expressed as:

$$\dot{\mathbf{v}} = \frac{cb}{m}\mathbf{u} + \mathbf{g}(\mathbf{r}); \ \frac{cb}{m} \equiv \Gamma \tag{2.5}$$

For the CSI case Eq. (2.5) is solved using the fact that c is constant as follows:

$$d\mathbf{v} = \frac{cb}{m}\mathbf{u}\,dt + \mathbf{g}(\mathbf{r})\,dt \tag{2.6}$$

Using Eq. (2.2),

$$d\mathbf{v} = -c\,\mathbf{u}\frac{dm}{m} + \mathbf{g}(\mathbf{r})\,dt \tag{2.7}$$

This can be integrated (assuming constant **u**) to yield:

$$\Delta\mathbf{v} = \mathbf{v}(t_f) - \mathbf{v}(t_o) = -c\mathbf{u}(\ln m_f - \ln m_o) + \int_{t_o}^{t_f} \mathbf{g}(\mathbf{r})\,dt \tag{2.8}$$

$$\Delta\mathbf{v} = c\,\mathbf{u}\,\ln\left(\frac{m_o}{m_f}\right) + \int_{t_o}^{t_f} \mathbf{g}(\mathbf{r})\,dt \tag{2.9}$$

which correctly indicates that, in the absence of gravity, the velocity change would be in the thrust direction **u**. The actual velocity change achieved also depends on the gravitational acceleration $\mathbf{g}(\mathbf{r})$ which is acting during the thrust period. The term in Eq. (2.9) involving the gravitational acceleration $\mathbf{g}(\mathbf{r})$ is called the *gravity loss*. Note there is no gravity loss due to an impulsive thrust.

If one ignores the gravity loss term for the time being, a cost functional representing propellant consumed can be formulated. As will be seen, minimizing this cost functional is equivalent to maximizing the final mass of the vehicle. Utilizing the fact that the thrust is equal to the product of the mass flow rate b and the exhaust velocity c, one can write:

$$\dot{m} = -b = \frac{-m\Gamma}{c} \tag{2.10}$$

$$\frac{dm}{m} = -\frac{\Gamma}{c}dt \tag{2.11}$$

For the CSI case the exhaust velocity c is constant and Eq. (2.11) can be integrated to yield

$$\ln\left(\frac{m_f}{m_o}\right) = -\frac{1}{c}\int_{t_o}^{t_f} \Gamma\, dt \tag{2.12}$$

or

$$c\ln\left(\frac{m_o}{m_f}\right) = \int_{t_o}^{t_f} \Gamma\, dt \equiv J_{CSI} \tag{2.13}$$

J_{CSI} is referred to as the *characteristic velocity* of the maneuver or the ΔV (pronounced "delta vee") and it is clear from Eq. (2.13) that minimizing J_{CSI} is equivalent to maximizing the final mass m_f.

In the impulsive thrust approximation for the unbounded thrust case ($\Gamma_{\max} \to \infty$) the vector thrust acceleration is represented by

$$\boldsymbol{\Gamma}(t) = \sum_{k=1}^{n} \Delta\mathbf{v}_k \delta(t - t_k) \tag{2.14}$$

with $t_o \leq t_1 < t_2 < \cdots < t_n \leq t_f$ representing the times of the n thrust impulses. (See Sections 6.1–3 of Ref. [2.1].) Using the definition of a unit impulse,

$$\int_{t_k^-}^{t_k^+} \delta(t - t_k)\, dt = 1 \tag{2.15}$$

and $t_k^{\pm} \equiv \lim \varepsilon \to 0(t_k \pm \varepsilon); \varepsilon > 0$.

Using Eq. (2.14) in Eq. (2.13):

$$J_{CSI} = \int_{t_o}^{t_f} \Gamma dt = \sum_{k=1}^{n} \Delta v_k \tag{2.16}$$

and the total propellant cost is given by the sum of the magnitudes of the velocity changes.

The corresponding cost functional for the VSI case is obtained differently. The exhaust power (stream or beam power) is half of the product of the thrust and the exhaust velocity:

$$P = \frac{1}{2}Tc = \frac{1}{2}m\Gamma c = \frac{1}{2}bc^2 \tag{2.17}$$

Using this along with

$$\frac{b}{m^2} = \frac{-\dot{m}}{m^2} = \frac{d}{dt}\left(\frac{1}{m}\right) \tag{2.18}$$

results in

$$\frac{d}{dt}\left(\frac{1}{m}\right) = \frac{\Gamma^2}{2P} \tag{2.19}$$

which integrates to

$$\frac{1}{m_f} - \frac{1}{m_o} = \frac{1}{2}\int_{t_o}^{t_f} \frac{\Gamma^2}{P}\,dt \tag{2.20}$$

Maximizing m_f for a given value of m_o regardless of whether Γ is optimal or not is obtained by running the engine at maximum power $P = P_{max}$. This is not as obvious as it looks in Eq. (2.20), because the value of Γ might be different for different values of P. To see that the engine should be run at maximum power note that for a specified trajectory $\mathbf{r}(t)$ the required vector thrust acceleration is given by Eq. (2.1) as

$$\mathbf{\Gamma}(t) = \ddot{\mathbf{r}}(t) - \mathbf{g}[\mathbf{r}(t)] \tag{2.21}$$

Thus, for a given trajectory $\mathbf{r}(t)$ (optimal or not), the final mass in Eq. (2.20) is maximized by running the engine at maximum power.

For this reason the VSI cost functional can be taken to be

$$J_{VSI} = \frac{1}{2}\int_{t_o}^{t_f} \Gamma^2\,dt \tag{2.22}$$

To summarize, the cost functionals representing minimum propellant expenditure are given by

$$J_{CSI} = \int_{t_o}^{t_f} \Gamma\,dt \tag{2.23}$$

and

$$J_{VSI} = \frac{1}{2}\int_{t_o}^{t_f} \Gamma^2\,dt \tag{2.24}$$

As seen in Eqs. (2.23) and (2.24) the minimum-propellant cost can be written in terms of the control $\Gamma(t)$ rather than introducing the mass as an additional state variable whose final value is to be maximized.

Problems

2.1 For an impulsive thrust consider first-order changes in $J = \Delta \text{v}$ and in the final mass m_f. Obtain an expression for the marginal (fractional) change $\delta m_f / m_f$ in terms of the marginal change $\delta J / J$. Assume the values of m_o and c are specified and there is no gravity loss.

2.2 Consider a constant thrust $T = T_o$ in a CSI propulsion system.

a) Determine an expression for $\dot{\Gamma}$ in terms of (only) the exhaust velocity c and the thrust acceleration magnitude Γ.

b) Solve the ODE in part (a) for $\Gamma(t)$. Express your answer in terms of T_o divided by a time-varying mass.

2.3 Show how Eq. (2.19) follows from Eq. (2.18).

Reference

[2.1] Prussing, J.E. and Conway, B.A., *Orbital Mechanics*, Oxford University Press, 2nd Edition, 2012.

3 Optimal Control Theory

3.1 Equation of motion and cost functional

Consider a dynamic system that operates between a specified initial time t_o and a final time t_f, which may be specified or unspecified. The equation of motion of the system can be written as

$$\dot{\mathbf{x}} = \mathbf{f}(\mathbf{x}, \mathbf{u}, t); \quad \mathbf{x}(t_o) = \mathbf{x}_o \quad \text{(specified)} \tag{3.1}$$

where $\mathbf{x}(t)$ is an $n \times 1$ *state vector* and $\mathbf{u}(t)$ is an $m \times 1$ *control vector*. Given an initial state $\mathbf{x}(t_o)$, the state $\mathbf{x}(t)$ is determined by the control $\mathbf{u}(t)$ through the equation of motion (3.1).

The first-order form of Eq. (3.1) is completely general because any higher-order equation can be transformed into first-order form. As an example consider Eq. (2.1):

$$\ddot{\mathbf{r}} = \mathbf{g}(\mathbf{r}) + \mathbf{\Gamma} \tag{2.1}$$

To transform to first-order form let

$$\dot{\mathbf{r}} = \mathbf{v} \tag{3.2a}$$

$$\dot{\mathbf{v}} = \mathbf{g} + \mathbf{\Gamma} \tag{3.2b}$$

Then

$$\dot{\mathbf{x}} = \begin{bmatrix} \dot{\mathbf{r}} \\ \dot{\mathbf{v}} \end{bmatrix} = \begin{bmatrix} \mathbf{v} \\ \mathbf{g} + \mathbf{\Gamma} \end{bmatrix} \tag{3.3}$$

and a second-order equation for a 3-vector $\mathbf{r}$ has been transformed into a first-order equation for a 6-vector $\mathbf{x}$.

Optimal Spacecraft Trajectories. John E. Prussing.

To formulate an optimal control problem we define a cost functional

$$J = \phi[\mathbf{x}(t_f), t_f] + \int_{t_o}^{t_f} L[\mathbf{x}(t), \mathbf{u}(t), t]\, dt \tag{3.4}$$

to be minimized by choice of a control $\mathbf{u}(t)$. The first term in Eq. (3.4) is sometimes called the "terminal cost" and the second term the "path cost".

As a simple example let $\phi = t_f$ and $L = 0$. Then the cost is $J = t_f$ and we have a minimum time problem.

As noted elsewhere, if we wish to maximize a performance index $\tilde{J}$, we minimize $J = -\tilde{J}$.

There are three basic forms of the cost functional, named after pioneers in the Calculus of Variations:

Mayer form: $J = \phi[\mathbf{x}(t_f), t_f]$

Lagrange form: $J = \int_{t_o}^{t_f} L(\mathbf{x}, \mathbf{u}, t)\, dt$

Bolza form: $J = \phi[\mathbf{x}(t_f), t_f] + \int_{t_o}^{t_f} L(\mathbf{x}, \mathbf{u}, t)\, dt$ [Eq. (3.4)]

There is also a special case of the Mayer form: $J = x_n(t_f)$, i.e., the cost is the final value of the last component of the state vector.

Although the Bolza form appears to be the most general, all the forms have equal generality. To see this we will transform the Bolza form into the special case of the Mayer form.

Example 3.1 Cost transformation

Given that $\mathbf{x}$ is an n-vector, define $\dot{x}_{n+1} = L$ with $x_{n+1}(t_o) = 0$. Then

$$\int_{t_o}^{t_f} L\, dt = x_{n+1}(t_f) - x_{n+1}(t_o) = x_{n+1}(t_f).$$

Then,

$$J = \phi[\mathbf{x}(t_f), t_f] + x_{n+1}(t_f) \equiv x_{n+2}(t_f) \tag{3.5}$$

3.2 General problem

Over the time interval $t_o \leq t \leq t_f$, where the final time can be specified or unspecified (fixed or free, closed or open) there is the dynamic constraint of the equation of motion:

$$\dot{\mathbf{x}} = \mathbf{f}(\mathbf{x}, \mathbf{u}, t); \quad \mathbf{x}(t_o) = \mathbf{x}_o \quad \text{(specified)} \tag{3.1}$$

and a cost to be minimized

$$J = \phi[\mathbf{x}(t_f), t_f] + \int_{t_o}^{t_f} L[\mathbf{x}(t), \mathbf{u}(t), t]\, dt \tag{3.4}$$

The ODE constraint of Eq. (3.1) is what makes Optimal Control Theory a generalization of the Calculus of Variations.

There may be *terminal constraints* $\boldsymbol{\psi}[\mathbf{x}(t_f), t_f] = \mathbf{0}$, for example, $\mathbf{r}(t_f) - \mathbf{r}^{\#}(t_f) = \mathbf{0}$, where $\mathbf{r}$ is the spacecraft position vector and $\mathbf{r}^{\#}$ is a target body position vector. This constraint then represents an orbital interception. And there may be *control constraints* such as a bounded control $|\mathbf{u}(t)| \leq M$.

First consider the simplest optimal control problem: fixed final time, no terminal or control constraints. We will derive first-order necessary conditions (NC) that must be satisfied to determine the control $\mathbf{u}^*(t)$ that minimizes the cost J. This is treated in more detail in Section 3.3 of Ref. [3.1] and Section 2.3 of Ref. [3.2].

Before we begin our derivation of the NC, a fundamental point needs to be made. We will be *assuming* that an optimal control $\mathbf{u}^*(t)$ exists and deriving conditions that must be satisfied. If no optimal control exists the NC are meaningless.

Peron's Paradox illustrates this point and goes as follows: Let $N =$ the largest integer. We will prove by contradiction that $N = 1$.

If $N > 1, N^2 > N$, which is a contradiction because N is the largest integer. Therefore, $N = 1$, a nonsensical result because there is no largest integer.

Assuming an optimal control $\mathbf{u}^*(t)$ exists for the general problem, let's proceed. First, we adjoin the equation of motion constraint to the cost to form

$$\bar{J} = \phi[\mathbf{x}(t_f), t_f] + \int_{t_o}^{t_f} L(\mathbf{x}, \mathbf{u}, t) + \boldsymbol{\lambda}^T[\mathbf{f}(\mathbf{x}, \mathbf{u}, t) - \dot{\mathbf{x}}]\, dt \tag{3.6}$$

Where $\boldsymbol{\lambda}(t)$ is an $n \times 1$ vector variously called the "adjoint" vector, the "costate" vector, or a "time-varying Lagrange multiplier" vector. Note that if the constraint equation of motion (3.1) is satisfied, we have added zero to the cost, in the same spirit as the use of a Lagrange multiplier in Section 2.2.1.

We will form the variation in the cost $\delta\bar{J}$ and derive NC that make $\delta\bar{J} = 0$ due to a small variation in the control about the optimal control $\delta\mathbf{u}(t) = \mathbf{u}(t) - \mathbf{u}^*(t)$. Note that the variation is equal to the difference at a common value of time t and for this reason it is sometimes called a *contemporaneous* variation, for which $\delta t \equiv 0$.

For convenience define the *Hamiltonian*

$$H(\mathbf{x}, \mathbf{u}, \boldsymbol{\lambda}, t) \equiv L(\mathbf{x}, \mathbf{u}, t) + \boldsymbol{\lambda}^T(t)\mathbf{f}(\mathbf{x}, \mathbf{u}, t) \tag{3.7}$$

(The terminology comes from classical mechanics, in which $H = \mathbf{p}^T\dot{\mathbf{q}} - L$, where L is the Lagrangian, $\mathbf{p}$ are the generalized momenta, and $\dot{\mathbf{q}}$ are the generalized velocities.)

Using the definition of the Hamiltonian, $\bar{J}$ in Eq. (3.6) is equal to $\bar{J} = \phi + \int_{t_o}^{t_f} [H - \lambda^T\dot{\mathbf{x}}]\, dt$.

Before forming the variation $\delta\bar{J}$ we will deal with the last term in $\bar{J}$, namely $-\int_{t_o}^{t_f} \boldsymbol{\lambda}^T\dot{\mathbf{x}}\, dt$.

First, we derive a vector integration by parts formula by using $d(\boldsymbol{\lambda}^T\mathbf{x}) = \boldsymbol{\lambda}^T d\mathbf{x} + d\boldsymbol{\lambda}^T\mathbf{x}$. Then $\int d(\boldsymbol{\lambda}^T\mathbf{x}) = \boldsymbol{\lambda}^T\mathbf{x} = \int \boldsymbol{\lambda}^T d\mathbf{x} + \int d\boldsymbol{\lambda}^T\mathbf{x}$. Using $d\mathbf{x} = \dot{\mathbf{x}}dt$ and $d\boldsymbol{\lambda}^T = \dot{\boldsymbol{\lambda}}^T dt$ we obtain $\int \boldsymbol{\lambda}^T\dot{\mathbf{x}}dt = \boldsymbol{\lambda}^T\mathbf{x} - \int \dot{\boldsymbol{\lambda}}^T\mathbf{x}dt$.

Applying this to the definite integral term $-\int_{t_o}^{t_f} \boldsymbol{\lambda}^T\dot{\mathbf{x}}\, dt$.

$$-\int_{t_o}^{t_f} \boldsymbol{\lambda}^T\dot{\mathbf{x}}dt = -\boldsymbol{\lambda}^T(t_f)\mathbf{x}(t_f) + \boldsymbol{\lambda}^T(t_o)\mathbf{x}(t_o) + \int_{t_o}^{t_f} \dot{\boldsymbol{\lambda}}^T(t)\mathbf{x}(t)\, dt \tag{3.8}$$

Using this result we rewrite Eq. (3.6) as

$$\bar{J} = \phi + \boldsymbol{\lambda}^T(t_o)\mathbf{x}(t_o) - \boldsymbol{\lambda}^T(t_f)\mathbf{x}(t_f) + \int_{t_o}^{t_f} [H + \dot{\boldsymbol{\lambda}}^T\mathbf{x}]dt \tag{3.9}$$

Forming the cost variation $\delta\bar{J}$ due to a control variation $\delta\mathbf{u}(t)$:

$$\begin{aligned}
\delta\bar{J} &= \delta\phi + \delta[\boldsymbol{\lambda}^T(t_o)\mathbf{x}(t_o)] - \delta[\boldsymbol{\lambda}^T(t_f)\mathbf{x}(t_f)] \\
&\quad + \delta\int_{t_o}^{t_f} [H + \dot{\boldsymbol{\lambda}}^T\mathbf{x}]\, dt \\
&= \frac{\partial\phi}{\partial\mathbf{x}(t_f)}\delta\mathbf{x}(t_f) + \boldsymbol{\lambda}^T(t_o)\delta\mathbf{x}(t_o) - \boldsymbol{\lambda}^T(t_f)\delta\mathbf{x}(t_f) \\
&\quad + \int_{t_o}^{t_f} \left[\frac{\partial H}{\partial\mathbf{x}}\delta\mathbf{x} + \frac{\partial H}{\partial\mathbf{u}}\delta\mathbf{u} + \dot{\boldsymbol{\lambda}}^T\delta\mathbf{x}\right] dt
\end{aligned} \tag{3.10}$$

Combining coefficients:

$$\delta\bar{J} = \left[\frac{\partial\phi}{\partial\mathbf{x}(t_f)} - \boldsymbol{\lambda}^T(t_f)\right]\delta\mathbf{x}(t_f) + \boldsymbol{\lambda}^T(t_o)\delta\mathbf{x}(t_o)$$
$$+ \int_{t_o}^{t_f}\left[\left(\frac{\partial H}{\partial\mathbf{x}} + \dot{\boldsymbol{\lambda}}^T\right)\delta\mathbf{x} + \frac{\partial H}{\partial\mathbf{u}}\delta\mathbf{u}\right]dt \tag{3.11}$$

Because $\delta\mathbf{x}(t)$ and $\delta\mathbf{u}(t)$ are arbitrary, for $\delta\bar{J} = 0$ the NC are obtained from Eq. (3.11) by setting individual coefficients equal to zero:

$$\dot{\boldsymbol{\lambda}}^T = -\frac{\partial H}{\partial\mathbf{x}} = -\frac{\partial L}{\partial\mathbf{x}} - \boldsymbol{\lambda}^T\frac{\partial\mathbf{f}}{\partial\mathbf{x}} \tag{3.12}$$

$$\boldsymbol{\lambda}^T(t_f) = \frac{\partial\phi}{\partial\mathbf{x}(t_f)} \tag{3.13}$$

$$\frac{\partial H}{\partial\mathbf{u}(t)} = \mathbf{0}^T \tag{3.14}$$

$$\text{either} \quad \delta x_k(t_o) = 0 \quad \text{or} \quad \lambda_k(t_o) = 0, \quad k = 1, 2, \ldots, n \tag{3.15}$$

where $\delta x_k(t_o)$ will be equal to 0 if the component $x_k(t_o)$ is specified, but if $x_k(t_o)$ is unspecified, $\lambda_k(t_o) = 0$. This represents a generalization of the initial condition $\mathbf{x}(t_o) = \mathbf{x}_o$ to allow some of the initial states to be unspecified. An additional NC is that the constraint Eq. (3.1) must be satisfied: $\dot{\mathbf{x}} = \mathbf{f}$ with initial conditions given by Eq. (3.15).

Equation (3.13) has a simple geometrical interpretation as a gradient vector. The final value of the adjoint vector is orthogonal to the line or surface ϕ = constant.

Note that there are $2n + m$ unknowns: $\mathbf{x}(t)$, $\boldsymbol{\lambda}(t)$, and $\mathbf{u}(t)$. Equation (3.1) with initial conditions given by (3.15) provide n equations; Eqs. (3.12) and (3.13) provide n equations; and Eq. (3.14) provides m equations, for a total of $2n + m$ NC equations in that same number of unknowns. So, in principle, the optimal control problem is solved! But there are several details to be worked out and observations to be made.

First note that even if the system equation $\dot{\mathbf{x}} = \mathbf{f}$ is nonlinear the adjoint equation (3.12) for $\boldsymbol{\lambda}(t)$ is *linear* because L and $\mathbf{f}$ are not functions of $\boldsymbol{\lambda}$. And if the cost is of the Mayer form (no L), the adjoint equation is homogeneous. But also note that the boundary condition (3.13) is at the *final* time in contrast to the conditions in Eq. (3.15) at the *initial* time. So, we have a split Two-Point Boundary Value Problem (2PBVP), which indicates a fundamental difficulty in solving the NC for an optimal control problem, because not all boundary conditions are given at the same value of time.

Note that the adjoint equation (3.12) and the system equation (3.1) are differential equations, but what is often called the *optimality condition* (3.14) is an algebraic equation that applies at all values of $t_o \leq t \leq t_f$. It requires that the optimal control result in a stationary value of H at each point along the trajectory. Later we will learn the stronger result that the optimal control must *minimize* the value of H.

A simple argument shows why Eq. (3.14) is an NC. Consider a scalar control u and assume $\partial H/\partial u \neq 0$ on a small finite interval Δt. Because u is arbitrary, if $\partial H/\partial u > 0$, choose $\delta u < 0$ and if $\partial H/\partial u < 0$, choose $\delta u > 0$. In either case $\partial H/\partial u\, \delta u < 0$, resulting in $\delta J < 0$ and a decrease in the cost. Therefore, it is necessary that $\partial H/\partial u = 0$ for J to be a minimum.

Hidden in Eq. (3.11) is an interpretation of the vector $\boldsymbol{\lambda}$. If all the other NC are satisfied

$$\delta \bar{J} = \delta J = \boldsymbol{\lambda}(t_o)^T \delta \mathbf{x}(t_o) \tag{3.16}$$

which provides the interpretation

$$\boldsymbol{\lambda}(t_o)^T = \frac{\partial J}{\partial \mathbf{x}(t_o)} \tag{3.17}$$

That is, each component $\lambda_k(t_o)$ is the sensitivity of the minimum cost to a small change in the corresponding component of the initial state. This means that the change in cost due to a small change $\delta x_k(t_o)$ can be calculated using Eq. (3.16) without resolving the problem.

Note that this is consistent with Eq. (3.15) that $\lambda_k(t_o) = 0$ if $x_k(t_o)$ is unspecified. The best value of $x_k(t_o)$ to minimize J is the value for which the cost is insensitive to small changes.

Finally, note that Eq. (3.17) can be generalized because any time t along an optimal trajectory can be considered to be the initial time for the remainder of the trajectory (as in today is the first day of the rest of your life). So, we can replace t_o by the time t and state that

$$\boldsymbol{\lambda}(t)^T = \frac{\partial J}{\partial \mathbf{x}(t)} \tag{3.18}$$

This is an application of Bellman's Principle of Optimality. Stated simply: Any part of an optimal trajectory is itself an optimal trajectory.

Example 3.2 Constant Hamiltonian

Form the total time derivative of $H(\mathbf{x}, \mathbf{u}, \boldsymbol{\lambda}, t)$:

$$\frac{dH}{dt} = \frac{\partial H}{\partial \mathbf{x}}\dot{\mathbf{x}} + \frac{\partial H}{\partial \mathbf{u}}\dot{\mathbf{u}} + \frac{\partial H}{\partial \boldsymbol{\lambda}}\dot{\boldsymbol{\lambda}} + \frac{\partial H}{\partial t} \tag{3.19}$$

Using $\dot{\mathbf{x}} = \mathbf{f}$ and $\partial H/\partial \boldsymbol{\lambda} = \mathbf{f}^T$

$$\frac{dH}{dt} = \left[\frac{\partial H}{\partial \mathbf{x}} + \dot{\boldsymbol{\lambda}}^T\right]\mathbf{f} + \frac{\partial H}{\partial \mathbf{u}}\dot{\mathbf{u}} + \frac{\partial H}{\partial t} \tag{3.20}$$

Applying the NC results in

$$\frac{dH}{dt} = \frac{\partial H}{\partial t} \tag{3.21}$$

So if $\partial H/\partial t = 0$, i.e., H is not an *explicit* function of t, then H is a constant. This occurs if neither L nor $\mathbf{f}$ is an explicit function of t, i.e., $L = L(\mathbf{x}, \mathbf{u})$ and $\mathbf{f} = \mathbf{f}(\mathbf{x}, \mathbf{u})$. That property for $\mathbf{f}$ occurs when the system is time invariant (autonomous). In classical mechanics the constant H is known as the *Jacobi Integral* and often represents conservation of total mechanical energy.

What good is a constant Hamiltonian? It can be used to evaluate the accuracy of a numerically integrated solution by monitoring the value of H as the solution is generated. Also, as with conservation of energy, if the value of H is known at one value of time, the unknown values of some variables at another time can be calculated.

Example 3.3 Rigid body control

Consider the single axis rotation of a rigid body for $0 \leq t \leq T$, where the final time T is specified. In terms of the angular velocity ω and the moment of inertia I the angular momentum is given by $h = I\omega$. With an external moment M_{ext} acting about the axis the equation of motion is $\dot{h} = I\dot{\omega} = M_{ext}$. In state and control variable notation, let $x = \omega$ and $u = M_{ext}/I$. Then the equation of motion is that of a simple integrator $\dot{x} = u$ with $x(0) = x_o$, a specified initial angular velocity.

To formulate an optimal control problem define a cost

$$J = \frac{1}{2}kx^2(T) + \frac{1}{2}\int_0^T u^2(t)dt \tag{3.22}$$

(The half factors are included to avoid factors of 2 when we differentiate.) The squared terms ensure that positive contributions to the cost occur regardless of the algebraic sign of the variables. The cost includes a terminal cost term to penalize the final state (angular velocity) and a path cost to penalize control (external moment). So, the effect is to slow down (for $k > 0$) the rotation without using an excessive amount of external moment, which presumably incurs some sort of cost such as propellant. The constant k is a weighting coefficient that determines the relative weight of each term in the cost. If we allow $k < 0$ we can represent a spin-up maneuver in which a large final angular velocity is desired. From the form of terminal cost term, we can expect that in the limit as $k \to \infty$ $x(T) \to 0$ and we stop the rotation. So, this example has the essential aspects of a simple optimal control problem for which $n = m = 1$.

To apply the NC form the Hamiltonian

$$H = \frac{1}{2}u^2 + \lambda u \tag{3.23}$$

Then the adjoint equation is

$$\dot{\lambda} = -\frac{\partial H}{\partial x} = 0 \tag{3.24}$$

so λ is constant and

$$\lambda(t) = \lambda(T) = \frac{\partial \phi}{\partial x(T)} = kx(T) \tag{3.25}$$

The optimality condition is

$$\frac{\partial H}{\partial u} = u + \lambda = 0 \tag{3.26}$$

so that

$$u = -\lambda = \text{constant} = -kx(T) \tag{3.27}$$

which indicates two important facts: (i) The adjoint variable λ provides important information to determine the optimal control, and (ii) of all the possible $u(t)$ for this problem the optimal control is a constant. (Note that, in addition to a stationary value of H, it is minimized because $H_{uu} = 1 > 0$.)

To solve the NC substitute for the control,

$$\dot{x} = -kx(T) \tag{3.28}$$

which yields

$$x^*(t) = -kx(T)t + x_o \tag{3.29}$$

Evaluating at $t = T$ and solving:

$$x^*(T) = \frac{x_o}{1 + kT} \tag{3.30}$$

and a larger $k > 0$ results in a smaller $x(T)$ because more weight is given to the terminal cost term.

The rest of the solution is given by

$$u^*(t) = \frac{-kx_o}{1 + kT} \tag{3.31}$$

and

$$x^*(t) = x_o \left[\frac{1 + k(T - t)}{1 + kT} \right] \tag{3.32}$$

where $x(t)$ is conveniently expressed in terms of the "time-to-go" $T - t$ and the state changes linearly in time from its initial value to its final value.

Some limiting cases are instructive: As $k \to \infty$, $x^*(T) \to 0$ and $u^*(t) \to -x_o/T$ indicating that as infinite weight is put on the terminal cost the control drives the initial state to zero in the allotted time.

As $k \to 0$, $x^*(T) \to x_o$, and $u^*(t) \to 0$ indicating that when all the weight is put on the control cost no control is used and the state remains unchanged.

The optimal cost is calculated to be

$$J^* = \frac{kx_o^2}{2(1 + kT)} \tag{3.33}$$

and we can explicitly verify the interpretation of $\lambda(0)$ in Eq. (3.17) as $\partial J^*/\partial x_o = kx_o/(1 + kT) = -u(0) = \lambda(0)$.

Finally, since $\partial H/\partial t = 0$, H = constant = $H(T)$ and

$$H(T) = -\frac{1}{2} \frac{k^2 x_o^2}{(1 + kT)^2} < 0 \tag{3.34}$$

And we note anecdotally that $\partial J^*/\partial T = H(T)$ (we will see why this is true later). This means that there is no optimal value of T; the larger the value of T the smaller the value of J^*.

With that observation we seem to have extracted all we can from this example!

3.3 Terminal constraints and unspecified final time

In addition to the general problem considered in Section 3.2 we now include an unspecified (open) final time t_f and terminal constraints of the form $\boldsymbol{\psi}[\mathbf{x}(t_f), t_f] = \mathbf{0}$, where $\boldsymbol{\psi}$ is a $q \times 1$ vector and $0 \leq q \leq n$. Our augmented cost is now

$$\bar{J} = \phi + \boldsymbol{\nu}^T \boldsymbol{\psi} + \int_{t_o}^{t_f} L + \boldsymbol{\lambda}^T[\mathbf{f} - \dot{\mathbf{x}}]dt \tag{3.35}$$

which is Eq. (3.6) with an additional terminal cost term and an unspecified upper limit on the integral. The vector $\boldsymbol{\nu}$ is a *constant* Lagrange multiplier q-vector.

For convenience define

$$\Phi[\mathbf{x}(t_f), t_f] \equiv \phi + \boldsymbol{\nu}^T \boldsymbol{\psi} \tag{3.36}$$

where the vector $\boldsymbol{\nu}$ has a useful interpretation (see Problem 3.4).

Consider the perturbation $d\bar{J}$ due to $\delta u(t)$, where the symbol "d" includes the effect of a differential change dt_f, for example, the *noncontemporaneous* variation $d\mathbf{x}(t_f) \equiv \mathbf{x}(t_f + dt_f) - \mathbf{x}^*(t_f)$. The result is a modified version of Eq. (3.11):

$$\begin{aligned} d\bar{J} &= \frac{\partial \Phi}{\partial \mathbf{x}(t_f)} d\mathbf{x}(t_f) + \left[\frac{\partial \Phi}{\partial t_f} + L_f\right] dt_f - \boldsymbol{\lambda}^T(t_f)\delta\mathbf{x}(t_f) + \boldsymbol{\lambda}^T(t_o)\delta\mathbf{x}(t_o) \\ &+ \int_{t_o}^{t_f} \left[\left(\frac{\partial H}{\partial \mathbf{x}} + \dot{\boldsymbol{\lambda}}^T\right)\delta\mathbf{x} + \frac{\partial H}{\partial \mathbf{u}}\delta\mathbf{u}\right] dt \end{aligned} \tag{3.37}$$

where $L_f \equiv L[(\mathbf{x}(t_f)), \mathbf{u}(t_f), t_f]$.

Next, we observe that $d\mathbf{x}(t_f) = \mathbf{x}(t_f + dt_f) - \mathbf{x}^*(t_f)$ is to first order equal to $\mathbf{x}(t_f) + \dot{\mathbf{x}}(t_f)dt_f - \mathbf{x}^*(t_f)$. Using $\delta\mathbf{x}(t_f) = \mathbf{x}(t_f) - \mathbf{x}^*(t_f)$ we can write $d\mathbf{x}(t_f) = \delta\mathbf{x}(t_f) + \dot{\mathbf{x}}(t_f)dt_f$, where the last term is the contribution due to the change dt_f. However,

$$\begin{aligned} \dot{\mathbf{x}}(t_f)dt_f &= [\dot{\mathbf{x}} * (t_f) + \delta\dot{\mathbf{x}}(t_f)]dt_f \\ &= \dot{\mathbf{x}} * (t_f) + dt_f + \text{second order term} \end{aligned} \tag{3.38}$$

So to first order we have the "$d - \delta$" rule:

$$d\mathbf{x}(t_f) = \delta\mathbf{x}(t_f) + \dot{\mathbf{x}} * (t_f)dt_f \tag{3.39}$$

This is shown in Figure 3.1.

Next use $\delta\mathbf{x}(t_f) = d\mathbf{x}(t_f) - \dot{\mathbf{x}}(t_f)dt_f$ to write Eq. (3.37) as

$$\begin{aligned} d\bar{J} &= \left[\frac{\partial \Phi}{\partial \mathbf{x}(t_f)} - \boldsymbol{\lambda}^T(t_f)\right] d\mathbf{x}(t_f) + \left[\frac{\partial \Phi}{\partial t_f} + L_f + \boldsymbol{\lambda}^T(t_f)\dot{\mathbf{x}}(t_f)\right] dt_f + \boldsymbol{\lambda}^T(t_o)\delta\mathbf{x}(t_o) \\ &+ \int_{t_o}^{t_f} \left[\left(\frac{\partial H}{\partial \mathbf{x}} + \dot{\lambda}^T\right)\delta\mathbf{x} + \frac{\partial H}{\partial \mathbf{u}}\delta\mathbf{u}\right] dt \end{aligned} \tag{3.40}$$

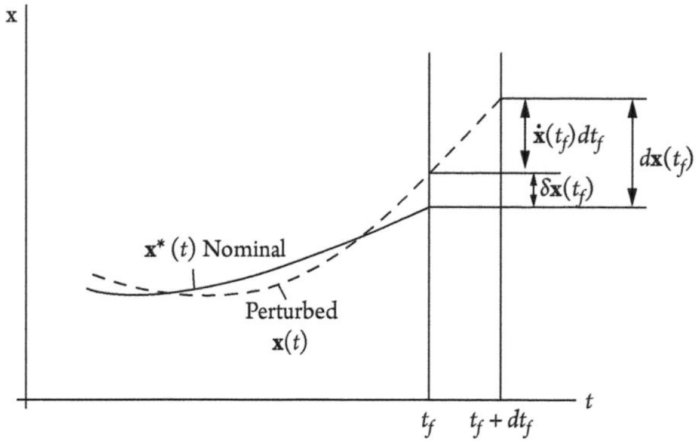

Figure 3.1 Relationship between $dx(t_f)$, $\forall x(t_f)$, and dt_f

3.4 Pontryagin minimum principle

The statement that the optimal control must provide a stationary value of the Hamiltonian, as expressed by the NC $\partial H/\partial \mathbf{u} = \mathbf{0}^T$ can be strengthened to the Minimum Principle, the NC that the optimal control must minimize the Hamiltonian, specifically,

$$H[\mathbf{x}^*, \mathbf{u}^*, \boldsymbol{\lambda}, t] \leq H[\mathbf{x}^*, \mathbf{u}, \boldsymbol{\lambda}, t] \tag{3.41}$$

This has great generality and applies even in the case that the control is an element of a closed set: $\mathbf{u}(t) \in U$. The standard reference for the Minimum Principle is Ref. [3.3] and Ref. [3.4] provides a short introduction to the subject.

The proof of this principle is notoriously difficult, but applying it is easy. A very simple example is for the bounded control: $-1 \leq u(t) \leq 1$ and the Hamiltonian $H = 3u$. Application of the principle determines $u^*(t) = -1$. In fact, for this example there is no value of u that provides a stationary value of H, so that weaker condition does not provide the optimal control.

Example 3.4 Minimum principle application

Consider a *minimum time* problem for a vehicle traveling at a *constant* scalar speed v_o. The control vector is a steering command in the form of a unit vector $\mathbf{u}$ in the direction of the velocity. The equation of motion is for $0 \leq t \leq T$:

$$\dot{\mathbf{r}} = v_o \mathbf{u}$$

with $\mathbf{r}(0) = \mathbf{r}_o \neq \mathbf{0}$ (specified). The terminal constraint is $\boldsymbol{\psi} = \mathbf{r}(T) = \mathbf{0}$. The initial time is zero and the cost is T.

Considering $\phi = T$ and $L = 0$ the Hamiltonian is $H = L + \boldsymbol{\lambda}^T\mathbf{f} = v_o\boldsymbol{\lambda}^T\mathbf{u}$.

Because H is linear in $\mathbf{u}$ it has no stationary value, so we choose $\mathbf{u}$ to minimize H by aligning $\mathbf{u}$ opposite to $\boldsymbol{\lambda}$ (to minimize the dot product) and enforce the fact that $\mathbf{u}$ is a unit vector. The result is $\mathbf{u} = -\boldsymbol{\lambda}/\lambda$.

Another NC is $\dot{\boldsymbol{\lambda}}^T = -\partial H/\partial \mathbf{r} = \mathbf{0}^T$ indicating that $\boldsymbol{\lambda}(t)$ is constant and hence $\mathbf{u}(t)$ is constant indicating a constant steering direction.

From Eq. (3.36)

$$\Phi = \phi + \boldsymbol{\nu}^T \boldsymbol{\psi} = T + \boldsymbol{\nu}^T \mathbf{r}(T)$$

and the boundary condition is simply $\boldsymbol{\lambda}^T(T) = \partial\Phi/\partial\mathbf{r}(T) = \boldsymbol{\nu}^T$. Using Eq. (3.47) we calculate $\Omega = 1 + \mathrm{v}_o \boldsymbol{\lambda}^T \mathbf{u} = 1 - \mathrm{v}_o \lambda = 0$ which yields the value of $\lambda = 1/\mathrm{v}_o$.

From the equation of motion $\dot{\mathbf{r}} = \mathrm{v}_o \mathbf{u}$, and because $\mathbf{u}$ is constant, we have $\mathbf{r}(t) - \mathbf{r}(0) = \mathrm{v}_o t \mathbf{u}$. Evaluating at the final time and applying the terminal constraint yields $\mathbf{r}(T) = \mathbf{r}_o + \mathrm{v}_o T \mathbf{u} = \mathbf{0}$, from which we solve for $\mathbf{u} = -\mathbf{r}_o/T\mathrm{v}_o$. Requiring $\mathbf{u}$ to be a unit vector yields the result that the minimum time is $T = r_o/\mathrm{v}_o$ and the optimal control is $\mathbf{u} = -\mathbf{r}_o/r_o$. Note that $\partial J/\partial r_o = \lambda_o = 1/\mathrm{v}_o$.

Finally note that this minimum-time example is actually a minimum-distance example made to sound more exciting as a minimum-time vehicle motion problem. The result is the well-known straight-line solution.

Problems

3.1 Based on the discussion at the end of the Preface, is the cost in Mayer form a function or a functional? Explain your answer.

3.2 Show that Eqs. (3.1) and (3.12) can be written as the pair of equations $\dot{\mathbf{x}} = \left(\frac{\partial H}{\partial \boldsymbol{\lambda}}\right)^T$ and $\dot{\boldsymbol{\lambda}} = -\left(\frac{\partial H}{\partial \mathbf{x}}\right)^T$, which are known as Hamilton's Equations in classical mechanics.

3.3 In Eq. (3.10) why are there no $\delta\boldsymbol{\lambda}$ terms?

3.4 The terminal constraint (3.45) can be generalized to $\boldsymbol{\psi}[\mathbf{x}(t_f), t_f, \mathbf{c}] = \mathbf{0}$, where $\mathbf{c}$ is a vector of constants in the constraints. Show that the vector $\boldsymbol{\nu}$ provides the interpretation of a change in the cost due to a small change in $\mathbf{c}$, namely, $\partial J/\partial \mathbf{c} = \boldsymbol{\nu}^T \boldsymbol{\psi}_{\mathbf{c}}$. Also show that in component form we have $\partial J/\partial c_i = \sum_{j=1}^{q} \nu_j \partial\psi_j/\partial c_i$.

3.5 Consider the *state variational equation* $\delta\dot{\mathbf{x}} = \frac{\partial \mathbf{f}}{\partial \mathbf{x}}\delta\mathbf{x} + \frac{\partial \mathbf{f}}{\partial \mathbf{u}}\delta\mathbf{u}$. Show for a cost in the Mayer form that along a solution satisfying the NC the product $\boldsymbol{\lambda}^T(t)\delta\mathbf{x}(t)$ is constant.

3.6 In Example 3.3 for $k < 0$:

a) Determine the range of values of k for which the solution makes sense.

b) Determine the maximum value of $x^*(T)$ that can be achieved.

3.7 a) In Example 3.4 consider $\mathbf{r} = [x\ y\ z]$ with specified initial position components x_o and y_o, but an unspecified value z_o. Determine the optimal solution including the optimal value of z_o *using the NC.*

b) Determine the values of $H(T)$ and Ω.

c)* Determine the optimal solution including the optimal value of α for the terminal constraint $\mathbf{r}(T) = \alpha\mathbf{a}$, where $\mathbf{a}$ is a specified vector for which $\mathbf{r}_o \times \mathbf{a} \neq \mathbf{0}$ and $-\infty < \alpha < \infty$ (sketch this).

References

[3.1] Longuski, J.M., Guzman, J.J., and Prussing, J.E., *Optimal Control with Aerospace Applications*, Springer, New York, 2014.

[3.2] Bryson, A.E. Jr and Ho, Y-C, *Applied Optimal Control*, Hemisphere Publishing, Washington, D.C., 1975.

[3.3] Pontryagin, L.S., Boltyanski, V.G., Gamkrelidze, R.V, and Mishchenko, E.F., *The Mathematical Theory of Optimal Processes*. Wiley, New York, 1962.

[3.4] Ross, I.M., *A Primer on Pontryagin's Principle in Optimal Control*, Collegiate Publishers, Carmel CA, 2009.

4 Optimal Trajectories

4.1 Optimal constant-specific-impulse trajectory

Constant-specific-impulse and variable-specific-impulse rocket engines are discussed in Section 2.3, with the cost functional for a CSI engine given by Eq. (2.23):

$$J_{CSI} = \int_{t_o}^{t_f} \Gamma \, dt \tag{4.1}$$

The equation of motion in first-order form is given by Eq. (3.2):

$$\dot{\mathbf{x}} = \begin{bmatrix} \dot{\mathbf{r}} \\ \dot{\mathbf{v}} \end{bmatrix} = \begin{bmatrix} \mathbf{v} \\ \mathbf{g} + \boldsymbol{\Gamma} \end{bmatrix} \tag{4.2}$$

The first-order necessary conditions for an optimal CSI trajectory were first derived by Lawden in Ref. [4.1] using classical Calculus of Variations. In the derivation that follows, an Optimal Control Theory formulation is used, but the derivation is similar to that of Lawden. One difference is that the mass is not defined as a state variable, but is kept track of indirectly.

In order to minimize the cost in Eq. (4.1) one forms the Hamiltonian using Eqs. (3.7) and (3.2) as

$$H = \Gamma + \boldsymbol{\lambda}_r^T \mathbf{v} + \boldsymbol{\lambda}_v^T [\mathbf{g}(\mathbf{r}) + \Gamma \mathbf{u}] \tag{4.3}$$

where we have written the thrust acceleration vector $\boldsymbol{\Gamma}$ as the product $\Gamma \mathbf{u}$, where Γ is the scalar thrust acceleration magnitude and $\mathbf{u}$ is a unit vector in the thrust direction.

The adjoint equations (3.41) are then

$$\dot{\boldsymbol{\lambda}}_r^T = -\frac{\partial H}{\partial \mathbf{r}} = -\boldsymbol{\lambda}_v^T \mathbf{G}(\mathbf{r}) \tag{4.4}$$

$$\dot{\boldsymbol{\lambda}}_v^T = -\frac{\partial H}{\partial \mathbf{v}} = -\boldsymbol{\lambda}_r^T \tag{4.5}$$

where

$$\mathbf{G}(\mathbf{r}) \equiv \frac{\partial \mathbf{g}(\mathbf{r})}{\partial \mathbf{r}} \tag{4.6a}$$

is the symmetric 3 × 3 *gravity gradient matrix*. (See Problem 4.1.)

Example 4.1 Derivation of a gravity gradient matrix

For the inverse-square gravitational field: $\mathbf{g}(\mathbf{r}) = -\frac{\mu}{r^2}\frac{\mathbf{r}}{r} = -\frac{\mu}{r^3}\mathbf{r}$ show that the gravity gradient matrix $\mathbf{G}(\mathbf{r})$ of Eq. (4.6) is equal to $\mathbf{G}(\mathbf{r}) = -\frac{\mu}{r^5}(3\mathbf{r}\mathbf{r}^T - r^2\mathbf{I}_3)$, where $\mathbf{I}_3$ is the 3 × 3 identity matrix.

For $\mathbf{g} = -\frac{\mu\mathbf{r}}{r^3}$ we have:

$$\mathbf{G} = \frac{\partial \mathbf{g}}{\partial \mathbf{r}} = \left[r^3\left(-\mu\frac{\partial \mathbf{r}}{\partial \mathbf{r}}\right) + \mu\mathbf{r}\left(3r^2\frac{\partial r}{\partial \mathbf{r}}\right)\right]/r^6 \tag{4.6b}$$

Using $\partial\mathbf{r}/\partial\mathbf{r} = \mathbf{I}_3$ and differentiating $r^2 = \mathbf{r}^T\mathbf{r}$ yields $2r(\partial r/\partial\mathbf{r}) = 2\mathbf{r}^T$, so $\partial r/\partial\mathbf{r} = \mathbf{r}^T/r$ and we obtain the final result:

$$\mathbf{G} = \frac{3\mu(r^2\mathbf{r}\mathbf{r}^T/r - \mu r^3\mathbf{I}_3)}{r^6} = \frac{\mu}{r^5}(3\mathbf{r}\mathbf{r}^T - r^2\mathbf{I}_3) \tag{4.6c}$$

For terminal constraints of the form

$$\boldsymbol{\psi}\left[\mathbf{r}(t_f), \mathbf{v}(t_f), t_f\right] = \mathbf{0} \tag{4.7}$$

which may describe an orbital intercept, rendezvous, etc., the boundary conditions on Eqs. (4.4–4.5) are given in terms of

$$\boldsymbol{\Phi} \equiv \boldsymbol{\nu}^T\boldsymbol{\psi}\left[\mathbf{r}(t_f), \mathbf{v}(t_f), t_f\right] \tag{4.8}$$

by Eq. (3.42):

$$\boldsymbol{\lambda}_r^T(t_f) = \frac{\partial \boldsymbol{\Phi}}{\partial \mathbf{r}(t_f)} = \boldsymbol{\nu}^T\frac{\partial \boldsymbol{\psi}}{\partial \mathbf{r}(t_f)} \tag{4.9}$$

$$\boldsymbol{\lambda}_v^T(t_f) = \frac{\partial \boldsymbol{\Phi}}{\partial \mathbf{v}(t_f)} = \boldsymbol{\nu}^T\frac{\partial \boldsymbol{\psi}}{\partial \mathbf{v}(t_f)} \tag{4.10}$$

There are two control variables, the thrust direction $\mathbf{u}$ and the thrust acceleration magnitude Γ, that must be chosen to satisfy the Minimum Principle discussed in Section 3.4, i.e., to minimize the instantaneous value of the Hamiltonian H. By inspection, the Hamiltonian of Eq. (4.3) is minimized over the choice of thrust direction by aligning the unit vector $\mathbf{u}(t)$ opposite to the adjoint vector $\boldsymbol{\lambda}_v(t)$. Because of the significance of the vector $-\boldsymbol{\lambda}_v(t)$ Lawden termed it the *primer vector* $\mathbf{p}(t)$:

$$\mathbf{p}(t) \equiv -\boldsymbol{\lambda}_v(t) \tag{4.11}$$

The optimal thrust unit vector is then in the direction of the primer vector, specifically:

$$\mathbf{u}(t) = \frac{\mathbf{p}(t)}{p(t)} \tag{4.12}$$

and

$$\boldsymbol{\lambda}_v^T \mathbf{u} = -\lambda_v = -p \tag{4.13}$$

in the Hamiltonian of Eq. (4.3).

From Eqs. (4.5) and (4.11) it is evident that

$$\boldsymbol{\lambda}_r(t) = \dot{\mathbf{p}}(t) \tag{4.14}$$

Equations (4.4, 4.5, 4.11, 4.14) combine to yield the *primer vector equation*

$$\ddot{\mathbf{p}} = \mathbf{G}(\mathbf{r})\mathbf{p} \tag{4.15}$$

The boundary conditions on the solution to Eq. (4.15) are obtained from Eqs. (4.9–4.10):

$$\mathbf{p}(t_f) = -\boldsymbol{\nu}^T \frac{\partial \boldsymbol{\psi}}{\partial \mathbf{v}(t_f)} \tag{4.16}$$

$$\dot{\mathbf{p}}(t_f) = \boldsymbol{\nu}^T \frac{\partial \boldsymbol{\psi}}{\partial \mathbf{r}(t_f)} \tag{4.17}$$

Note that in Eq. (4.16) the final value of the primer vector for an optimal intercept is the zero vector, because the terminal constraint $\boldsymbol{\psi}$ does not depend on $\mathbf{v}(t_f)$.

Using Eqs. (4.11–4.15) the Hamiltonian of Eq. (4.3) can be rewritten as

$$H = -(p-1)\Gamma + \dot{\mathbf{p}}^T\mathbf{v} - \mathbf{p}^T\mathbf{g} \tag{4.18}$$

To minimize the Hamiltonian over the choice of the thrust acceleration magnitude Γ, one notes that the Hamiltonian is a linear function of Γ and thus the minimizing

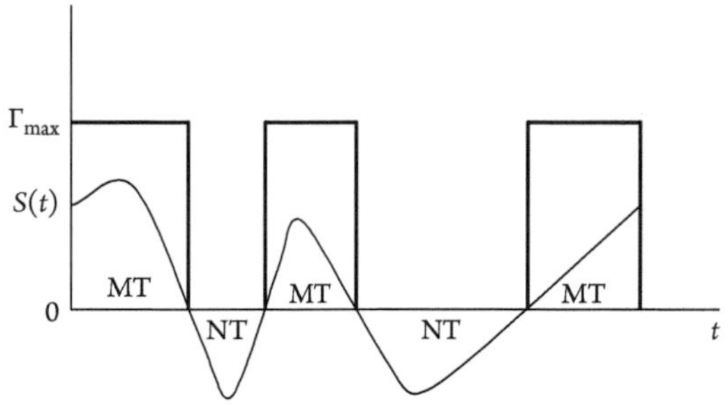

Figure 4.1 Three-burn CSI switching function and thrust profile.

value for $0 \le \Gamma \le \Gamma_{max}$ will depend on the algebraic sign of the coefficient of Γ in Eq. (4.18). It is convenient to define the *switching function*

$$S(t) \equiv p - 1 \tag{4.19}$$

The choice of the thrust acceleration magnitude Γ that minimizes H is then given by the "bang-bang" control law:

$$\Gamma = \begin{cases} \Gamma_{max} & \text{for} \quad S > 0 \ (p > 1) \\ 0 & \text{for} \quad S < 0 \ (p < 1) \end{cases} \tag{4.20}$$

That is, the thrust magnitude switches between its limiting values of 0 (an NT *null-thrust* arc) and T_{max} (an MT *maximum-thrust* arc) each time $S(t)$ passes through 0 ($p(t)$ passes through 1) according to Eq. (4.20). Figure 4.1 shows an example switching function for a three-burn trajectory.

The possibility also exists that $S(t) \equiv 0$ ($p(t) \equiv 1$) on an interval of finite duration. From Eq. (4.18) it is evident that in this case the thrust acceleration magnitude is not determined by the Minimum Principle and may take on intermediate values between 0 and Γ_{max}. This IT "intermediate thrust arc" in Ref. [4.1] is referred to as a *singular arc* in optimal control (Ref. [4.2]).

Lawden explained the origin of the term *primer vector* in a personal letter in 1990: *In regard to the term 'primer vector', you are quite correct in your supposition. I served in the artillery during the war [World War II] and became familiar with the initiation of the burning of cordite by means of a primer charge. Thus, p = 1 is the signal for the rocket motor to be ignited.*

For a CSI engine running at constant maximum thrust T_{max} the mass flow rate b_{max} will be constant. If the engine is on for a total of Δt time units,

$$\Gamma_{max}(t) = T_{max}/(m_o - b_{max}\Delta t) \tag{4.21}$$

Other necessary conditions are that the variables $\mathbf{p}$ and $\dot{\mathbf{p}}$ must be continuous everywhere. Equation (4.19) then indicates that the switching function $S(t)$ is also continuous everywhere.

Even though the gravitational field is time-invariant, the Hamiltonian in this formulation does not provide a first integral (constant of the motion) on an MT arc, because Γ is an explicit function of time as shown in Eq. (4.21). From Eq. (4.18)

$$H = -S\Gamma + \dot{\mathbf{p}}^T\mathbf{v} - \mathbf{p}^T\mathbf{g} \tag{4.22}$$

Note that the Hamiltonian is continuous everywhere because $S = 0$ at the discontinuities in the thrust acceleration magnitude.

4.2 Optimal impulsive trajectory

For a high-thrust CSI engine the thrust durations are very small compared with the times between thrusts. Because of this one can approximate each MT arc as an impulse (Dirac delta function) having unbounded magnitude ($\Gamma_{max} \rightarrow \infty$) and zero duration. The primer vector then determines both the optimal times and directions of the thrust impulses with $p \leq 1$ corresponding to $S \leq 0$. The impulses can occur only at those instants at which $S = 0$ ($p = 1$). These impulses are separated by NT arcs along which $S < 0$ ($p < 1$). At the impulse times the primer vector is then a unit vector in the optimal thrust direction.

The necessary conditions (NC) for an optimal impulsive trajectory first derived by Lawden are in Ref. [4.1].

For a linear system these NC are also sufficient conditions (SC) for an optimal trajectory, Appendix C and (Ref. [4.3]). Also in Appendix C, an upper bound on the number of impulses required for an optimal solution is given.

Figure 4.2 shows a trajectory (at top) and a primer vector magnitude (at bottom) for an optimal three-impulse solution. The orbital motion is counterclockwise and canonical units are used. The canonical time unit is the orbital period of the circular orbit that has a radius of one canonical length unit. The initial orbit is a unit radius circular orbit shown as the topmost orbit going counterclockwise from the symbol ⊕ at (1, 0) to (−1, 0). The transfer time is 0.5 original (initial) orbit periods (OOP). The target is in a coplanar circular orbit of radius 2, with an initial lead angle (ila) of 270° and shown by the symbol □ at (0, −2). The spacecraft departs ○ and intercepts □ at approximately (1.8, −0.8) as shown. The + signs at the initial and final points indicate thrust impulses and the + sign on the transfer orbit very near (0, 0) indicates the location of the midcourse impulse. The magnitudes of the three Δvs are shown at the left, with the total Δv equal to 1.3681 in units of circular orbit speed in the initial orbit.

The bottom graph in Fig. 4.2 displays the time history of the primer vector magnitude. Note that it satisfies the necessary conditions of Table 4.1 for an optimal transfer.

The examples shown in this chapter are coplanar, but the theory and applications apply to three-dimensional trajectories as well, e.g., Prussing and Chiu, Ref. [4.4].

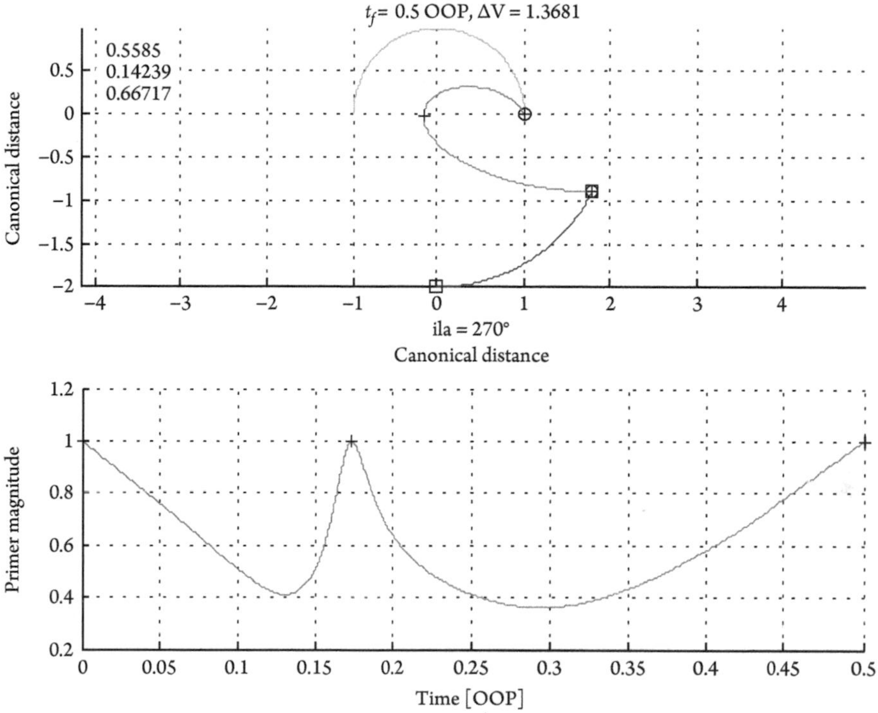

Figure 4.2 Optimal three-impulse trajectory and primer magnitude. Figure 4.2 was generated using the MATLAB computer code written by S.L. Sandrik for Ref. [4.9].

Table 4.1 Impulsive necessary conditions

1. The primer vector and its first derivative are continuous everywhere.
2. The magnitude of the primer vector satisfies $p(t) \leq 1$ with the impulses occurring at those instants at which $p = 1$.
3. At the impulse times the primer vector is a unit vector in the optimal thrust direction.
4. As a consequence of the above conditions, $dp/dt = \dot{p} = \dot{\mathbf{p}}^T\mathbf{p} = 0$ at an intermediate impulse (not at the initial or final time).

4.3 Optimal variable-specific-impulse trajectory

A variable specific impulse (VSI) engine is also known as a power-limited (PL) engine, because the power source is separate from the engine itself, e.g., solar panels, radioisotope thermoelectric generator (RTG), etc. The power delivered to the engine is bounded between 0 and a maximum value P_{max}, with the optimal value being constant and equal to the maximum, as discussed in Section 2.3. The cost functional representing minimum propellant consumption for the VSI case is given by Eq. (2.22) as

$$J_{VSI} = \frac{1}{2}\int_{t_o}^{t_f} \Gamma^2(t)\,dt \tag{4.23}$$

Writing Γ^2 as $\boldsymbol{\Gamma}^T\boldsymbol{\Gamma}$ the corresponding Hamiltonian function can be written as

$$H = \frac{1}{2}\boldsymbol{\Gamma}^T\boldsymbol{\Gamma} + \boldsymbol{\lambda}_r^T\mathbf{v} + \boldsymbol{\lambda}_v^T[\mathbf{g}(\mathbf{r}) + \boldsymbol{\Gamma}] \tag{4.24}$$

For the VSI case there is no need to consider the thrust acceleration magnitude and direction separately, so the vector $\boldsymbol{\Gamma}$ is used in place of the term $\Gamma\mathbf{u}$ that appears in Eq. (4.3).

Because H is a nonlinear function of $\boldsymbol{\Gamma}$ the Minimum Principle is applied by setting

$$\frac{\partial H}{\partial \boldsymbol{\Gamma}} = \boldsymbol{\Gamma}^T + \boldsymbol{\lambda}_v^T = \mathbf{0}^T \tag{4.25}$$

or

$$\boldsymbol{\Gamma}(t) = -\boldsymbol{\lambda}_v(t) = \mathbf{p}(t) \tag{4.26}$$

using the definition of the primer vector in Eq. (4.11). Thus, for a VSI engine the optimal thrust acceleration vector is equal to the primer vector: $\boldsymbol{\Gamma}(t) = \mathbf{p}(t)$.

Because of this Eq. (4.2), written as $\ddot{\mathbf{r}} = \mathbf{g}(\mathbf{r}) + \boldsymbol{\Gamma}$, can be combined with Eq. (4.15), as in Ref. [4.5] to yield a fourth-order differential equation in $\mathbf{r}$:

$$\mathbf{r}^{iv} - \dot{\mathbf{G}}\dot{\mathbf{r}} + \mathbf{G}(\mathbf{g} - 2\ddot{\mathbf{r}}) = \mathbf{0} \tag{4.27}$$

Every solution to Eq. (4.27) is an optimal VSI trajectory through the gravity field $\mathbf{g}(\mathbf{r})$. But desired boundary conditions, such as specified position and velocity vectors at the initial and final times, must be satisfied.

From (Eq. 4.25) we get:

$$\frac{\partial^2 H}{\partial \Gamma^2} = \frac{\partial}{\partial \Gamma}\left(\frac{\partial H}{\partial \Gamma}\right)^T = \mathrm{I}_3 \tag{4.28}$$

where $\mathbf{I}_3$ is the 3×3 identity matrix. Eq. (4.28) shows that the (Hessian) matrix of second partial derivatives is positive definite, verifying that H is minimized.

Because the VSI thrust acceleration of Eq. (4.26) is continuous, a procedure described in Ref. [4.6] to test whether second-order NC and SC are satisfied can be applied. Equation (4.28) shows that a NC for minimum cost (Hessian matrix positive semi-definite) and part of the SC (Hessian matrix positive definite) are satisfied. The other condition that is both a NC and SC is the Jacobi no-conjugate-point condition. Reference [4.6] details the test for that.

4.4 Solution to the primer vector equation

The primer vector equation Eq. (4.15) can be written in first-order form as the linear system:

$$\frac{d}{dt}\begin{bmatrix}\mathbf{p}\\ \dot{\mathbf{p}}\end{bmatrix} = \begin{bmatrix}\mathbf{O}_3 & \mathbf{I}_3\\ \mathbf{G} & \mathbf{O}_3\end{bmatrix}\begin{bmatrix}\mathbf{p}\\ \dot{\mathbf{p}}\end{bmatrix} \tag{4.29}$$

where $\mathbf{O}_3$ is the 3×3 zero matrix.

Equation (4.29) is of the form $\dot{\mathbf{y}} = \mathbf{F}(t)\,\mathbf{y}$ and its solution can be written in terms of a transition matrix $\mathbf{\Phi}(t, t_o)$ as

$$\mathbf{y}(t) = \mathbf{\Phi}(t, t_o)\mathbf{y}(t_o) \tag{4.30}$$

for a specified initial condition $\mathbf{y}(t_o)$.

Glandorf in Ref. [4.7] and Battin in Ref. [4.8] present forms of the transition matrix for an inverse-square gravitational field. [In Ref. [4.7] the missing Eq. (2.33) is $\Phi(t, t_o) = P(t)P^{-1}(t_o)$.]

Note that on an NT (no-thrust or coast) arc the variational (linearized) state equation is, from Eq. (4.2)

$$\delta\dot{\mathbf{x}} = \begin{bmatrix}\delta\dot{\mathbf{r}}\\ \delta\dot{\mathbf{v}}\end{bmatrix} = \begin{bmatrix}\mathbf{O}_3 & \mathbf{I}_3\\ \mathbf{G} & \mathbf{O}_3\end{bmatrix}\begin{bmatrix}\delta\mathbf{r}\\ \delta\mathbf{v}\end{bmatrix} \tag{4.31}$$

which is the same equation as Eq. (4.29). So, the transition matrix in Eq. (4.30) is also the transition matrix for the state variation, i.e., the *state transition matrix* in Ref. [4.8].

Also shown in Ref. [4.8] is that this state transition matrix has the usual properties from linear system theory (Appendix D) and is also symplectic, the definition of which is that

$$\mathbf{\Phi}^T(t, t_o)\mathbf{J}\mathbf{\Phi}(t, t_o) = \mathbf{J} \tag{4.32}$$

where

$$\mathbf{J} = \begin{bmatrix}\mathbf{O}_3 & \mathbf{I}_3\\ -\mathbf{I}_3 & \mathbf{O}_3\end{bmatrix} \tag{4.33}$$

Note that $\mathbf{J}^2 = -\mathbf{I}_6$ indicating that $\mathbf{J}$ is a matrix analog of the imaginary number $i = \sqrt{-1}$.

As shown in Problem 4.7 the inverse of the $\mathbf{\Phi}$ matrix can be simply obtained by matrix multiplication. This is especially useful when the state transition matrix is determined numerically because the inverse matrix $\mathbf{\Phi}^{-1}(t, t_o) = \mathbf{\Phi}(t_o, t)$ can be computed without explicitly inverting a 6×6 matrix. As seen from Eq. (4.30) this inverse matrix yields the variable $\mathbf{y}(t_o)$ in terms of $\mathbf{y}(t)$: $\mathbf{y}(t_o) = \mathbf{\Phi}(t_o, t)\mathbf{y}(t)$.

4.5 A vector constant

Consider the vector

$$\mathbf{A} = \mathbf{p} \times \mathbf{v} - \dot{\mathbf{p}} \times \mathbf{r} \tag{4.34}$$

On an NT arc in an arbitrary gravitational field

$$\dot{\mathbf{A}} = \dot{\mathbf{p}} \times \mathbf{v} + \mathbf{p} \times \dot{\mathbf{v}} - \ddot{\mathbf{p}} \times \mathbf{r} - \dot{\mathbf{p}} \times \dot{\mathbf{r}} \tag{4.35}$$

Using $\ddot{\mathbf{p}} = \mathbf{G}\mathbf{p}$, $\dot{\mathbf{r}} = \mathbf{v}$, and $\dot{\mathbf{v}} = \mathbf{g}$ we get

$$\dot{\mathbf{A}} = \mathbf{p} \times \mathbf{g} - \mathbf{G}\mathbf{p} \times \mathbf{r} \tag{4.36}$$

which is a zero vector for an inverse-square gravitational field (see Problem 4.8). Thus, **A** is a constant vector and provides useful information. It has a direct interpretation in optimal orbit transfer into or out of a circular terminal orbit (see Section 5.2.1).

At *any* optimal impulse (whether it is intermediate or terminal)

$$\mathbf{A}^+ - \mathbf{A}^- = \mathbf{p} \times (\mathbf{v}^+ - \mathbf{v}^-) - \dot{\mathbf{p}} \times (\mathbf{r}^+ - \mathbf{r}^-) \tag{4.37}$$

which, because **v** is discontinuous at an impulse, becomes

$$\mathbf{A}^+ - \mathbf{A}^- = \mathbf{p} \times \Delta\mathbf{v} = \Delta v(\mathbf{p} \times \mathbf{p}) = \mathbf{0} \tag{4.38}$$

and thus **A** is continuous at any optimal impulse.

Problems

4.1 Show that the gravity gradient matrix in Eq. (4.6) is symmetric by expressing $\mathbf{g}(\mathbf{r})$ in terms of a scalar potential function: $\mathbf{g}^T = -\partial U(\mathbf{r})/\partial \mathbf{r}$. Determine the expression for an element G_{ij}.

4.2 a) Derive Eq. (4.27).

b) Briefly outline a procedure for determining $\dot{\mathbf{G}}$.

4.3 Calculate the (total) time derivative of the Hamiltonian function along a CSI trajectory that satisfies the first-order necessary conditions (NC) in an arbitrary gravitational field. Consider the upper bound on the thrust magnitude $T_{\max}$ to be constant and allow for a time-varying mass.

a) Determine whether H is constant on NT, MT, and IT arcs.

b) Determine whether H is discontinuous at the switching times.

4.4 Calculate the (total) time derivative of the Hamiltonian along a VSI trajectory that satisfies the NC in an arbitrary gravitational field.

4.5 a) Show that $\mathbf{p}(t) = k_o\mathbf{v}(t)$ (where $\mathbf{v}$ is the velocity vector and k_o is an arbitrary scalar constant) is *a solution* to the primer vector equation on an NT arc in an arbitrary gravitational field.

b) Determine the initial conditions $\mathbf{p}_o$ and $\dot{\mathbf{p}}_o$ in terms of the initial conditions $\mathbf{r}_o$ and $\mathbf{v}_o$.

4.6 Determine whether the matrix $\mathbf{J}$ is itself symplectic.

4.7 Show that a symplectic matrix has the property $\mathbf{\Phi}^{-1} = -\mathbf{J}\mathbf{\Phi}^T\mathbf{J}$ and comment on the similarity to an orthogonal matrix.

4.8 Verify that $\dot{\mathbf{A}} = \mathbf{0}$ in Eq. (4.36) for an inverse-square gravitational field.

4.9* For a general linear system $\dot{\mathbf{y}} = \mathbf{F}\mathbf{y}$ determine the required condition on $\mathbf{F}$ that results in a symplectic transition matrix.

References

[4.1] Lawden, D.F., *Optimal Trajectories for Space Navigation*, Butterworths, London, 1963.

[4.2] Bryson, A.E. Jr. and Ho Y-C, *Applied Optimal Control*, Hemisphere Publishing Co., Washington D.C., 1975.

[4.3] Prussing, J.E., "Optimal impulsive linear systems: sufficient conditions and maximum number of impulses", *The Journal of the Astronautical Sciences*, Vol, 43, No. 2, Apr–Jun 1995, pp. 195–206.

[4.4] Prussing, J.E., and Chiu, J-H, "Optimal multiple-impulse time-fixed rendezvous between circular orbits", *Journal of Guidance, Control, and Dynamics*, Vol. 9. No. 1, Jan–Feb 1986, pp. 17–22. also *Errata*, Vol. 9, No. 2, p. 255.

[4.5] Prussing, J.E., "Equation for optimal power-limited spacecraft trajectories", *Journal of Guidance, Control, and Dynamics*, Vol. 16, No. 2, Mar–Apr 1993, pp. 391–3.

[4.6] Prussing, J.E., and Sandrik, S.L., "Second-order necessary conditions and sufficient conditions applied to continuous-thrust trajectories", *Journal of Guidance, Control, and Dynamics*, Vol 28, No. 4, Jul–Aug 2005, pp. 812–16.

[4.7] Glandorf, D.R., Lagrange multipliers and the state transition matrix for coasting arcs, *AIAA Journal*, Vol. 7, No. 2, Feb 1969, pp. 363–5.

[4.8] Battin, R.H., *An Introduction to the Mathematics and Methods of Astrodynamics*, Revised Edition, AIAA Education Series, New York, 1999, Section 9.7.

[4.9] Sandrik, S.L., *Primer-Optimized Results and Trends for Circular Phasing and Other Circle-to-Circle Impulsive Coplanar Rendezvous*, Ph.D. Thesis, 2006, University of Illinois at Urbana-Champaign.

[4.10] Pines, S., "Constants of the Motion for Optimum Thrust Trajectories in a Central Force Field", *AIAA Journal*, Vol. 2, No. 11, Nov 1964, pp. 2010–14.

5 Improving a Nonoptimal Impulsive Trajectory

5.1 Fixed-time-impulsive rendezvous

If the primer vector evaluated along an impulsive trajectory fails to satisfy the necessary conditions (NC) of Table 4.1 in Section 4.2 for an optimal solution, the way in which the NC are violated provides information that can lead to a solution that does satisfy the NC. The necessary gradients were first derived by Lion and Handelsman in Ref. [5.1]. and implemented in a nonlinear programming algorithm by Jezewski and Rozendaal in Ref. [5.2]. Reference [5.1] forms part of a body of knowledge known as Primer Vector Theory.

In this chapter a somewhat simplified derivation of the basic results of Ref. [5.1] is presented. In addition, numerous figures are included to illustrate the concepts and applications. In Section 5.2 the theory is applied to orbit transfer, as contrasted with orbital rendezvous.

For a given set of boundary conditions and transfer time an impulsive trajectory can be modified either by allowing a terminal coast or by adding a midcourse impulse. A terminal coast can be either an initial coast, in which the first impulse occurs after the initial time, or a final coast, in which the final impulse occurs before the final time. In the former case, the spacecraft coasts along the initial orbit after the initial time until the first impulse occurs. In the latter case the rendezvous actually occurs before the final time and the spacecraft coasts along the final orbit until the final time is reached.

To determine when a terminal coast will result in a trajectory that has a lower fuel cost, consider the two-impulse fixed-time rendezvous trajectory shown in Fig. 5.1.

In the two-body problem, if the terminal radii $\mathbf{r}_o$ and $\mathbf{r}_f$ are specified along with the transfer time $\tau \equiv t_f - t_o$, the solution to Lambert's Problem provides the terminal velocity vectors $\mathbf{v}_o^+$ and $\mathbf{v}_f^-$ on the transfer orbit. Because the velocity vectors are known on the initial orbit ($\mathbf{v}_o^-$ before the first impulse) and on the final orbit ($\mathbf{v}_f^+$ after the final

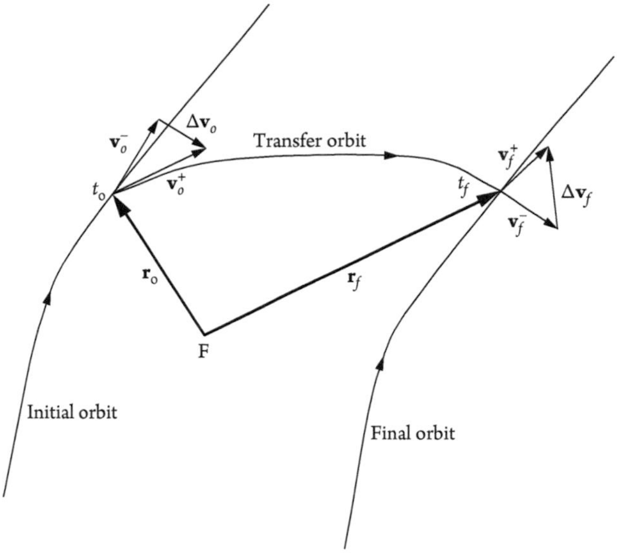

Figure 5.1 A fixed-time impulsive rendezvous trajectory

impulse), the required velocity changes can be determined as

$$\Delta\mathbf{v}_o = \mathbf{v}_o^+ - \mathbf{v}_o^- \tag{5.1}$$

and

$$\Delta\mathbf{v}_f = \mathbf{v}_f^+ - \mathbf{v}_f^- \tag{5.2}$$

Once the vector velocity changes are known, the primer vector can be evaluated along the trajectory to determine if the NC are satisfied. In order to satisfy the NC that on an optimal trajectory the primer vector at an impulse time is a unit vector in the direction of the impulse, one imposes the following boundary conditions on the primer vector:

$$\mathbf{p}(t_o) \equiv \mathbf{p}_o = \frac{\Delta\mathbf{v}_o}{\Delta v_o} \tag{5.3}$$

$$\mathbf{p}(t_f) \equiv \mathbf{p}_f = \frac{\Delta\mathbf{v}_f}{\Delta v_f} \tag{5.4}$$

The primer vector can then be evaluated along the transfer orbit using the 6×6 transition matrix solution

$$\begin{bmatrix} \mathbf{p}(t) \\ \dot{\mathbf{p}}(t) \end{bmatrix} = \mathbf{\Phi}(t, s) \begin{bmatrix} \mathbf{p}(s) \\ \dot{\mathbf{p}}(s) \end{bmatrix} \tag{5.5}$$

where the 3 × 3 partitions of the transition matrix are designated as

$$\mathbf{\Phi}(t,s) \equiv \begin{bmatrix} \mathbf{M}(t,s) & \mathbf{N}(t,s) \\ \mathbf{S}(t,s) & \mathbf{T}(t,s) \end{bmatrix} \tag{5.6}$$

Equation (5.6) can then be evaluated for the designated terminal times to yield

$$\mathbf{p}_f = \mathbf{M}_{fo}\mathbf{p}_o + \mathbf{N}_{fo}\dot{\mathbf{p}}_o \tag{5.7}$$

and

$$\dot{\mathbf{p}}_f = \mathbf{S}_{fo}\mathbf{p}_o + \mathbf{T}_{fo}\dot{\mathbf{p}}_o \tag{5.8}$$

where the abbreviated notation is used that $\mathbf{p}_f \equiv \mathbf{p}(t_f)$ and $\mathbf{M}_{fo} \equiv \mathbf{M}(t_f, t_o)$.

Equation (5.7) can be solved for the initial primer rate vector

$$\dot{\mathbf{p}}_o = \mathbf{N}_{fo}^{-1}\left[\mathbf{p}_f - \mathbf{M}_{fo}\mathbf{p}_o\right] \tag{5.9}$$

The matrix $\mathbf{N}_{fo}^{-1}$ exists except at isolated values of $\tau = t_f - t_o$. Assuming τ is not equal to one of these values, with both the primer vector and the primer vector rate known at the initial time, the primer vector along the transfer orbit for $t_o \le t \le t_f$ can be calculated as

$$\mathbf{p}(t) = \mathbf{N}_{to}\mathbf{N}_{fo}^{-1}\frac{\Delta\mathbf{v}_f}{\Delta v_f} + \left[\mathbf{M}_{to} - \mathbf{N}_{to}\mathbf{N}_{fo}^{-1}\mathbf{M}_{fo}\right]\frac{\Delta\mathbf{v}_o}{\Delta v_o} \tag{5.10}$$

5.1.1 Criterion for a terminal coast

One of the options available to modify a two-impulse solution that does not satisfy the NC for an optimal transfer is to include a terminal coast period, in the form of either an initial coast, a final coast, or both. As shown in Fig. 5.2, one allows the possibility that the initial impulse occurs at time $t_o + dt_o$ due to a coast in the initial orbit of duration $dt_o > 0$ and that the final impulse occurs at a time $t_f + dt_f$. In the case of a final coast $dt_f < 0$ in order that the final impulse occurs prior to the nominal final time allowing a coast in the final orbit until the nominal final time. However, for clarity in Fig. 5.2, the value of dt_f is shown as positive. A positive value also has a physical interpretation as we will see.

To determine whether a terminal coast will lower the cost of the trajectory, an expression for the difference in cost between the perturbed trajectory (with the terminal coasts) and the nominal trajectory (without the coasts) must be derived. The derivation that follows is a somewhat simplified version of the original by Lion and Handelsman in Ref. [5.1]. The cost on the nominal trajectory is simply

$$J = \Delta v_o + \Delta v_f \tag{5.13}$$

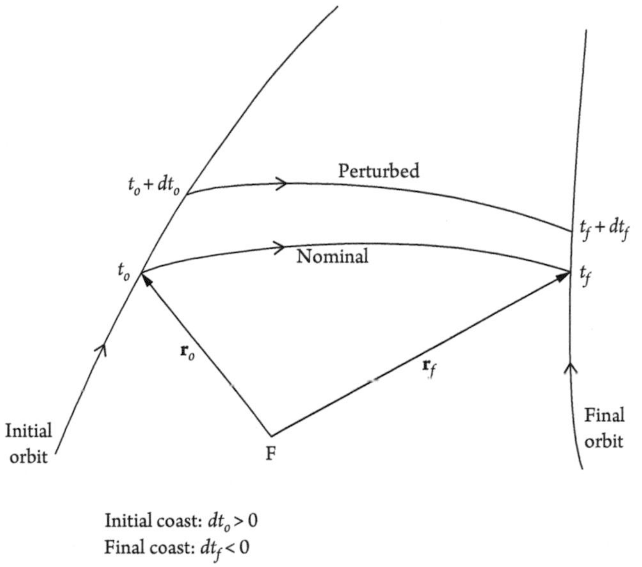

Figure 5.2 Terminal coasts of duration dt_o and dt_f

for the two-impulse solution. In order to determine the differential change in the cost due to the differential coast periods, the concept of a noncontemporaneous, or "skew" variation is needed. As explained prior to Eq. (3.39) this variation combines two effects: The variation due to being on a perturbed trajectory and the variation due to a difference in the time. The variable d denotes a noncontemporaneous variation in contrast to the variable δ that represents a contemporaneous variation. The rule for relating the two types of variations is given by Eq. (3.39) as

$$d\mathbf{x}(t_o) = \delta\mathbf{x}(t_o) + \dot{\mathbf{x}}_o^* dt_o \tag{5.14}$$

where the variation in the initial state has been used as an example.

The noncontemporaneous variation in the cost must be determined. Because the coast periods result in changes in the vector velocity changes, the variation in the cost can be expressed, from Eq. (5.13) as

$$dJ = \frac{\partial \Delta v_o}{\partial \Delta \mathbf{v}_o} d\Delta\mathbf{v}_o + \frac{\partial \Delta v_f}{\partial \Delta \mathbf{v}_f} d\Delta\mathbf{v}_f \tag{5.15}$$

As is explained prior to Eq. (4.6b)

$$\frac{\partial a}{\partial \mathbf{a}} = \frac{\mathbf{a}^T}{a} \tag{5.16}$$

Using the result of Eq. (5.16) the variation in the cost in Eq. (5.15) can be expressed as

$$dJ = \frac{\Delta \mathbf{v}_o^T}{\Delta v_o} d\Delta \mathbf{v}_o + \frac{\Delta \mathbf{v}_f^T}{\Delta v_f} d\Delta \mathbf{v}_f \tag{5.17}$$

Finally, Eq. (5.17) can be rewritten in terms of the initial and final primer vector using Eqs. (5.3–5.4) as

$$dJ = \mathbf{p}_o^T d\Delta \mathbf{v}_o + \mathbf{p}_f^T d\Delta \mathbf{v}_f \tag{5.18}$$

Next, one is faced with evaluating a noncontemporaneous variation in a velocity change, such as $d\Delta \mathbf{v}_o$. First, one notes that

$$d\Delta \mathbf{v}_o = d(\mathbf{v}_o^+ - \mathbf{v}_o^-) = d\mathbf{v}_o^+ - d\mathbf{v}_o^- \tag{5.19}$$

and that the variations in position and velocity on the initial orbit are

$$d\mathbf{r}_o = \mathbf{v}_o^- dt_o; \quad d\mathbf{v}_o^- = \dot{\mathbf{v}}_o^- dt_o \tag{5.20}$$

and the variation in velocity on the transfer orbit at the initial time is obtained using Eq. (5.14)

$$d\mathbf{v}_o^+ = \delta \mathbf{v}_o^+ + \dot{\mathbf{v}}_o^+ dt_o \tag{5.21}$$

However,

$$\dot{\mathbf{v}}_o^+ = \dot{\mathbf{v}}_o^- = \mathbf{g}(\mathbf{r}_o) \tag{5.22}$$

because the acceleration of the spacecraft is due only to gravity when the engine is off, and the gravitational acceleration $\mathbf{g}$ is a continuous function of the position. Thus Eq. (5.19) becomes

$$d\Delta \mathbf{v}_o = \delta \mathbf{v}_o^+ \tag{5.23}$$

By similar analysis one determines that

$$d\Delta \mathbf{v}_f = -\delta \mathbf{v}_f^- \tag{5.24}$$

and Eq. (5.18) for the variation in the cost can be expressed as

$$dJ = \mathbf{p}_o^T \delta \mathbf{v}_o^+ - \mathbf{p}_f^T \delta \mathbf{v}_f^- \tag{5.25}$$

This is not the final result because the contemporaneous variations in the velocity at the ends of the transfer orbit must be determined. Recall from Problem 3.5 that the product

$$\boldsymbol{\lambda}^T(t)\delta \mathbf{x}(t) = \text{constant} \tag{5.26}$$

between thrust impulses. From Eqs. (4.11) and (4.14) $\boldsymbol{\lambda}^T = \left[\boldsymbol{\lambda}_r^T \; \boldsymbol{\lambda}_v^T\right] = \left[\dot{\mathbf{p}}^T \; -\mathbf{p}^T\right]$. Thus Eq. (5.26) can be expressed as

$$\dot{\mathbf{p}}^T(t)\delta\mathbf{r}(t) - \mathbf{p}^T(t)\delta\mathbf{v}(t) = \text{constant} \tag{5.27}$$

Equating the values at the ends of the transfer orbit:

$$\dot{\mathbf{p}}_o^T\delta\mathbf{r}_o - \mathbf{p}_o^T\delta\mathbf{v}_o^+ = \dot{\mathbf{p}}_f^T\delta\mathbf{r}_f - \mathbf{p}_f^T\delta\mathbf{v}_f^- \tag{5.28}$$

By rearranging terms in Eq. (5.28) one can express the variation in cost of Eq. (5.25) as

$$dJ = \dot{\mathbf{p}}_o^T\delta\mathbf{r}_o - \dot{\mathbf{p}}_f^T\delta\mathbf{r}_f \tag{5.29}$$

The advantage of Eq. (5.29) is that the variations in the velocity vectors have been replaced by variations in the position vectors. These can be related to the differential times dt_o and dt_f as will be now be seen. From Eq. (5.14)

$$\delta\mathbf{r}_o = d\mathbf{r}_o - \mathbf{v}_o^+ dt_o \tag{5.30}$$

and, analogously

$$\delta\mathbf{r}_f = d\mathbf{r}_f - \mathbf{v}_f^- dt_f \tag{5.31}$$

Using the relationship in Eq. (5.20) that on the initial orbit $d\mathbf{r}_o = \mathbf{v}_o^- dt_o$ Eq. (5.30) becomes

$$\delta\mathbf{r}_o = -(\mathbf{v}_o^+ - \mathbf{v}_o^-)dt_o = -\Delta\mathbf{v}_o dt_o \tag{5.32}$$

and at the final time $d\mathbf{r}_f = \mathbf{v}_f^+ dt_f$ which leads to

$$\delta\mathbf{r}_f = \Delta\mathbf{v}_f dt_f \tag{5.33}$$

Figure 5.3 shows the interpretation of the vectors $\delta\mathbf{r}_o$ and $d\mathbf{r}_o$ as the position variation on the perturbed trajectory at the original initial time t_o and the position variation of the perturbed orbit at the (later) time $t_o + dt_o$, respectively. Also shown is the result in Eq. (5.32) that the vector $\delta\mathbf{r}_o$ has direction opposite to $\Delta\mathbf{v}_o$. The position variation $d\mathbf{r}_o$ is due to the initial coast along the initial orbit. Figure 5.4 shows the analogous situation at the final time. Note that the direction of $\delta\mathbf{r}_f$ is opposite to $\Delta\mathbf{v}_f$ because $dt_f < 0$.

Now Eq. (5.29) can be expressed as

$$dJ = -\dot{\mathbf{p}}_o^T\Delta\mathbf{v}_o dt_o - \dot{\mathbf{p}}_f^T\Delta\mathbf{v}_f dt_f \tag{5.34}$$

The final form of the expression for the variation in cost is obtained by expressing the vector velocity changes in terms of the primer vector using Eqs. (5.3–5.4) as

$$dJ = -\Delta v_o \dot{\mathbf{p}}_o^T\mathbf{p}_o dt_o - \Delta v_f \dot{\mathbf{p}}_f^T\mathbf{p}_f dt_f \tag{5.35}$$

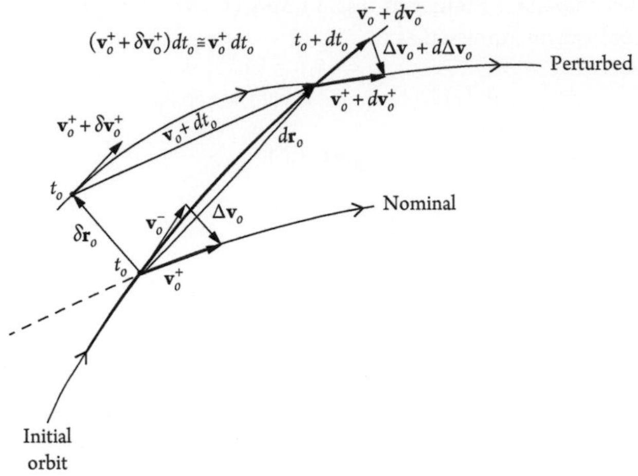

Figure 5.3 Interpretation of $\delta\mathbf{r}_o$ and $d\mathbf{r}_o$

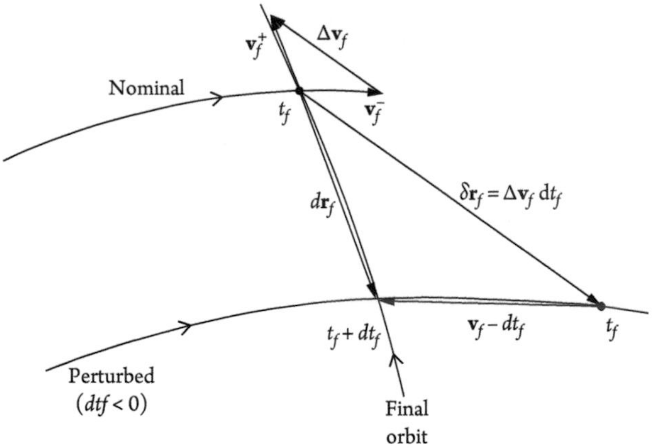

Figure 5.4 Interpretation of $\delta\mathbf{r}_f$ and $d\mathbf{r}_f$

In Eq. (5.35) one can identify the gradients of the cost with respect to the terminal impulse times as

$$\frac{\partial J}{\partial t_o} = -\Delta v_o \dot{\mathbf{p}}_o^T \mathbf{p}_o \tag{5.36}$$

and

$$\frac{\partial J}{\partial t_f} = -\Delta v_f \dot{\mathbf{p}}_f^T \mathbf{p}_f \tag{5.37}$$

One notes that the dot products in Eqs. (5.36–5.37) are simply the slopes of the primer magnitude time history at the terminal times due to the fact that $p^2 = \mathbf{p}^T\mathbf{p}$ and, after differentiation with respect to time, $2p\dot{p} = 2\dot{\mathbf{p}}^T\mathbf{p}$. Because $p = 1$ at the terminal times,

$$\frac{dp}{dt} = \dot{p} = \dot{\mathbf{p}}^T\mathbf{p} \tag{5.38}$$

The criteria for adding an initial or final coast can now be summarized.

If $\dot{p}_o > 0$ an initial coast (represented by $dt_o > 0$) will lower the cost.

Similarly, if $\dot{p}_f < 0$ a final coast (represented by $dt_f < 0$) will lower the cost.

It is worth noting that, conversely, if $\dot{p}_o \leq 0$ an initial coast will not lower the cost. This is consistent with the NC for an optimal solution and represents an alternate proof of the NC that $p \leq 1$ on an optimal solution. Similarly, if $\dot{p}_f \geq 0$ a final coast will not lower the cost. However, one can interpret these results even further. If $\dot{p}_o < 0$ a value of $dt_o < 0$ yields $dJ < 0$ indicating that an *earlier* initial impulse time will lower the cost. This is the opposite of an initial coast and simply means that the cost can be lowered by increasing the transfer time by starting the transfer earlier. Similarly, a value of $\dot{p}_f > 0$ implies that a $dt_f > 0$ will yield $dJ < 0$. In this case the cost can be lowered by increasing

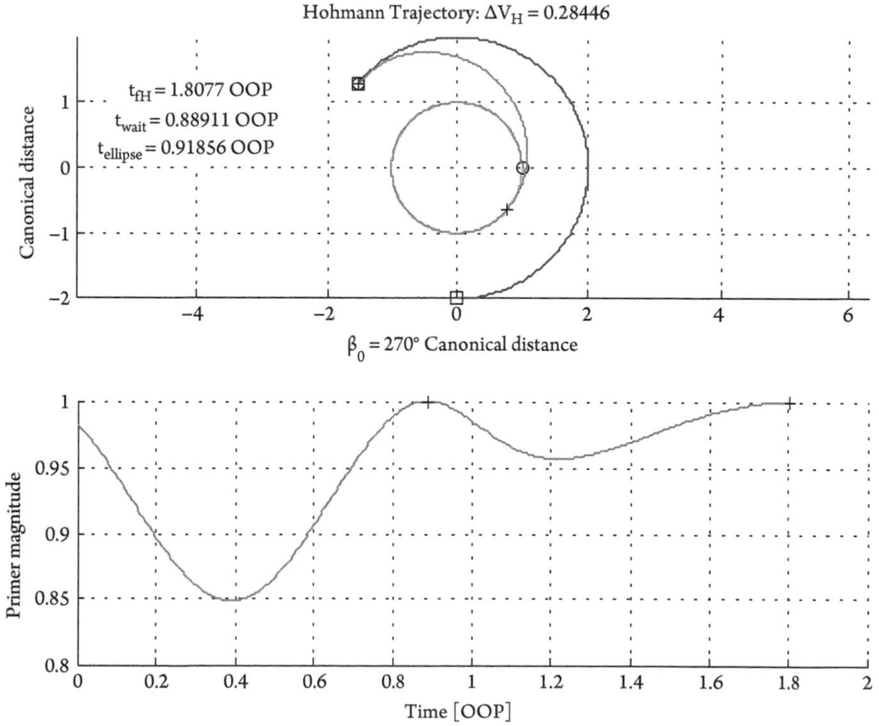

Figure 5.5 Hohmann transfer with initial coast

the transfer time by increasing the time of the final impulse. From these observations, one can conclude that for a time-open optimal solution, such as the Hohmann transfer, the slopes of the primer magnitude time history must be zero at the impulse times, indicating that no improvement in the cost can be made by adjusting the times of the impulses. Figure 5.5 illustrates this for a Hohmann transfer with an initial coast. In Fig. 5.5 the original orbit is a unit-radius circular orbit and the final orbit is a circular orbit of radius equal to 2. The direction of orbital motion is counter-clockwise and the unit of time is Original Orbit Period (OOP). The initial lead angle (ila) of the target is 270°, requiring a long initial coast of almost an entire OOP. The square symbols indicate the initial and final target locations. The circle indicates the initial vehicle location and the plus signs indicate the locations of the first and second impulses. The values of t_o, t_1, t_f are equal to 0, 0.88911, and 1.8077 OOP, respectively.

Figure 5.6 shows a primer history that violates the NC in a manner indicating that an initial cost will lower the cost. (That primer history also indicates that a final coast will lower the cost, but we will choose an initial coast. As shown by Sandrik in Ref. [5.5] choosing a final coast instead will usually lead to a different final result.) For an initial coast, the gradient of Eq. (5.36) is used in a nonlinear programming algorithm to iterate on the time of the first impulse. This is a one-dimensional search in which small finite changes in the time of the first impulse are incrementally made using the gradient of

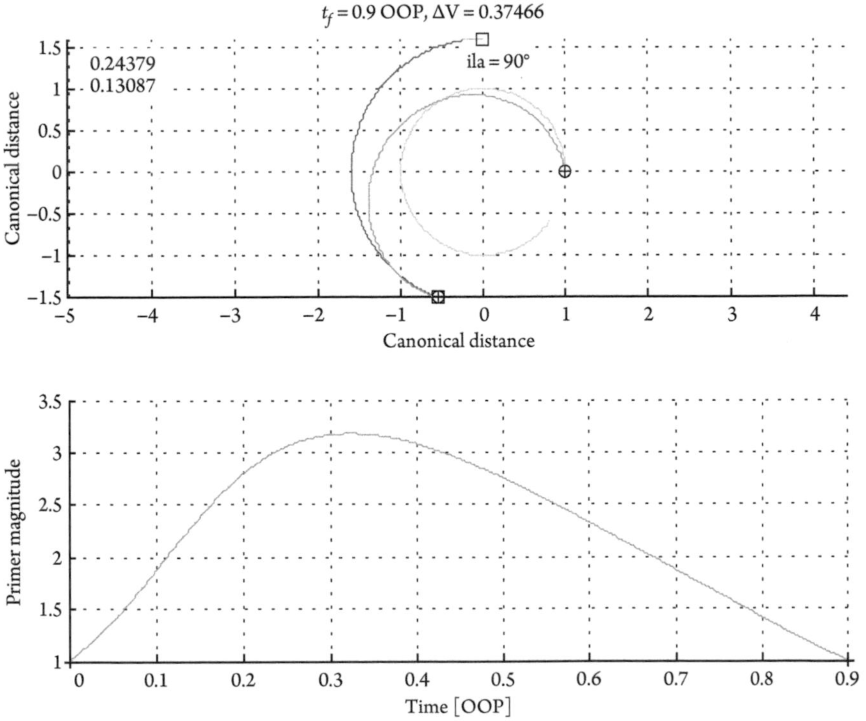

Figure 5.6 Primer magnitude indicating initial/final coast

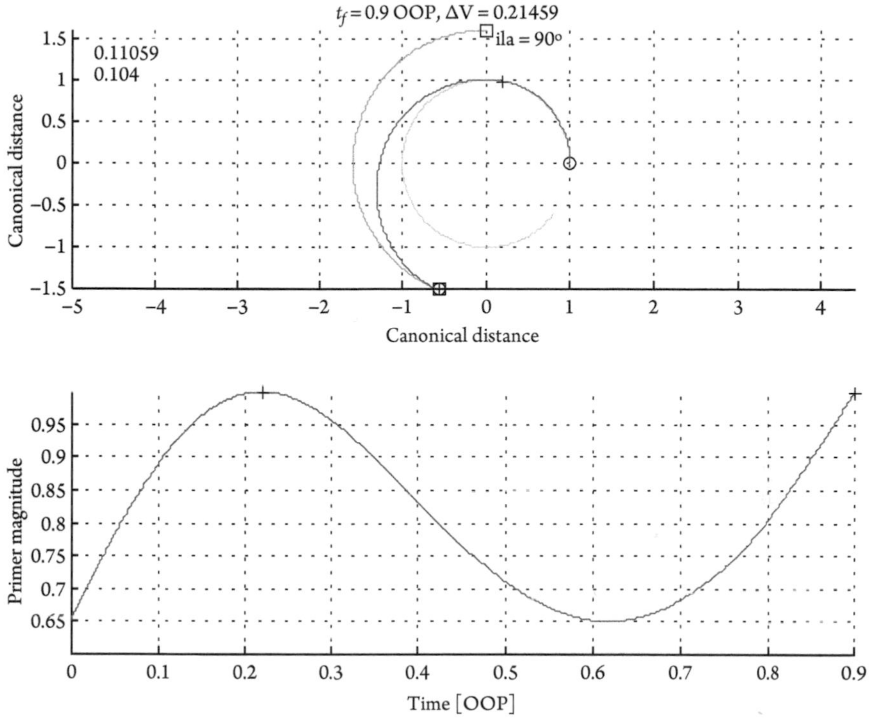

Figure 5.7 Optimal initial coast trajectory and primer magnitude

Eq. (5.36) until the gradient is driven to zero. On each iteration, new values for the terminal velocity changes are calculated and a new primer vector solution is obtained. Note that once the iteration begins, the time of the first impulse is no longer t_o, but a later value denoted by t_1. In a similar way, if the final impulse time becomes an iteration variable, it will be denoted by t_n, where the last impulse is considered to be the nth impulse. For a two-impulse trajectory, n is equal to two, but as will be seen shortly, optimal solutions can require more than two impulses. When the times of the first and last impulses become iteration variables, in all the formulas in the preceding analysis the subscript o is replaced by 1 everywhere and f is replaced by n.

Figures 5.5–7, 5.9, 5.10, and 5.12 were generated using the MATLAB computer code written by S.L. Sandrik for Ref. [5.5].

Figure 5.7 shows the converged result of an iteration on the time of the initial impulse. Note that the necessary condition $p \leq 1$ is satisfied and the gradient of the cost with respect to the t_1, the time of the first impulse is zero because $\dot{p}_1 = 0$ making the gradient of Eq. (5.36) equal to zero. This simply means that a small change in t_1 will cause no change in the cost, i.e., the cost has achieved a stationary value and satisfies the first-order necessary conditions.

Comparing Figs. 5.6 and 5.7 we see that the cost has decreased 43% from 0.37466 to 0.21459 and that for this solution an initial coast but no final coast is required.

5.1.2 Criterion for addition of a midcourse impulse

Besides terminal coasts, the addition of a midcourse impulse is another potential way of lowering the cost of a nonoptimal impulsive trajectory. The addition of an impulse is more complicated than including a terminal coast because four new parameters are introduced: Three components of the position of the impulse and the time of the impulse. One must derive a criterion that indicates when the addition of an impulse will lower the cost and then also determine where in space and when in time the impulse should be applied. It will turn out that the where and when will be done in two steps. The first step is to determine initial values of position and time of the new impulse that will lower the cost. The second step is to iterate on the position and time using gradients that will be developed, until a minimum of the cost is achieved. Note that this procedure is more complicated than for terminal coasts, because the starting value of the coast time for the iteration was simply taken to be zero, i.e., no coast.

When considering the addition of a midcourse impulse, let us assume $dt_o = dt_f = 0$, i.e., there are no terminal coasts. Because we are doing a first-order perturbation analysis, superposition applies and we can combine the previous results for terminal coasts easily with our new results. Also, we will discuss the case of adding a third impulse to a two-impulse trajectory, but the same theory applies to the case of adding a midcourse impulse to any two-impulse segment of an n-impulse trajectory.

The cost on the nominal, two-impulse trajectory is given by

$$J = \Delta v_o + \Delta v_f \tag{5.39}$$

The variation in the cost due to adding an impulse is given by adding the midcourse velocity change magnitude Δv_m to Eq. (5.15):

$$dJ = \frac{\partial \Delta v_o}{\partial \Delta \mathbf{v}_o} d\Delta \mathbf{v}_o + \Delta v_m + \frac{\partial \Delta v_f}{\partial \Delta \mathbf{v}_f} d\Delta \mathbf{v}_f \tag{5.40}$$

This simplifies using the same analysis as in Eq. (5.25) to yield

$$dJ = \mathbf{p}_o^T \delta \mathbf{v}_o^+ + \Delta v_m - \mathbf{p}_f^T \delta \mathbf{v}_f^- \tag{5.41}$$

One can next use Eq. (5.26) with the admonition that the value of the constant on the first segment, from the initial impulse to the midcourse impulse, need not be the same as the value on the second segment, from the midcourse impulse to the final impulse. Also, because dt_o and dt_f are zero, it follows that $\delta \mathbf{r}_o$ and $\delta \mathbf{r}_f$ are zero from Eqs. (5.32–5.33). Using Eq. (5.25) and equating the values at the ends of the first segment yields

$$-\mathbf{p}_o^T \delta \mathbf{v}_o^+ = \dot{\mathbf{p}}_m^T \delta \mathbf{r}_m - \mathbf{p}_m^T \delta \mathbf{v}_m^- \tag{5.42}$$

and on the second segment:

$$\dot{\mathbf{p}}_m^T \delta \mathbf{r}_m - \mathbf{p}_m^T \delta \mathbf{v}_m^+ = -\mathbf{p}_f^T \delta \mathbf{v}_f^- \tag{5.43}$$

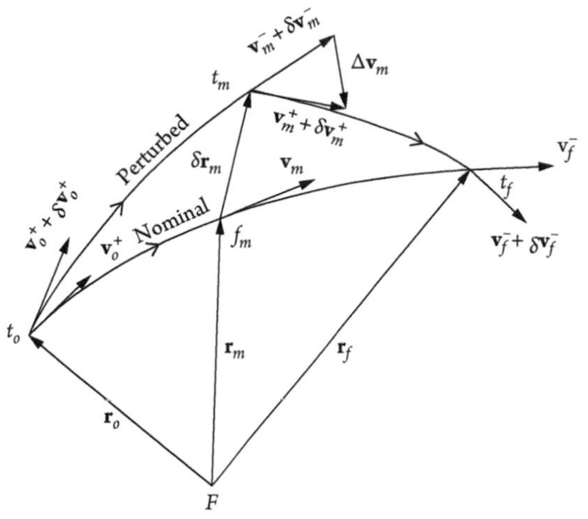

Figure 5.8 Nominal and perturbed trajectories

where $\delta\mathbf{v}_m^-$ and $\delta\mathbf{v}_m^+$ are the variations in the velocity vectors on the perturbed trajectory immediately before and after the midcourse impulse. The perturbed and nominal trajectories are shown in Fig. 5.8.

The midcourse velocity change is related to these velocity variations by

$$\Delta\mathbf{v}_m = (\mathbf{v}_m + \delta\mathbf{v}_m^+) - (\mathbf{v}_m + \delta\mathbf{v}_m^-) = \delta\mathbf{v}_m^+ - \delta\mathbf{v}_m^- \tag{5.44}$$

Substituting Eqs. (5.42) and (5.43) into the variation in the cost of Eq. (5.41) yields

$$\begin{aligned} dJ &= \mathbf{p}_m^T\delta\mathbf{v}_m^- + \Delta v_m - \mathbf{p}_m^T\delta\mathbf{v}_m^+ \\ &= \Delta v_m - \mathbf{p}_m^T(\delta\mathbf{v}_m^+ - \delta\mathbf{v}_m^-) \\ &= \Delta v_m\left(1 - \mathbf{p}_m^T\frac{\Delta\mathbf{v}_m}{\Delta v_m}\right) \end{aligned} \tag{5.45}$$

where Eq. (5.44) has been used.

In Eq. (5.45) the expression for dJ involves a dot product between the primer vector and a unit vector. If the numerical value of this dot product is greater than one, $dJ < 0$ and the perturbed trajectory has a lower cost than the nominal trajectory. In order for the value of the dot product to be greater than one, it is necessary that $p_m > 1$. Here again we have an alternative derivation of the necessary condition that $p \leq 1$ on an optimal trajectory. We also have the criterion that tells us when the addition of a midcourse impulse will lower the cost.

If the value of $p(t)$ exceeds unity along the trajectory, the addition of a mid-course impulse at a time for which $p > 1$ will lower the cost.

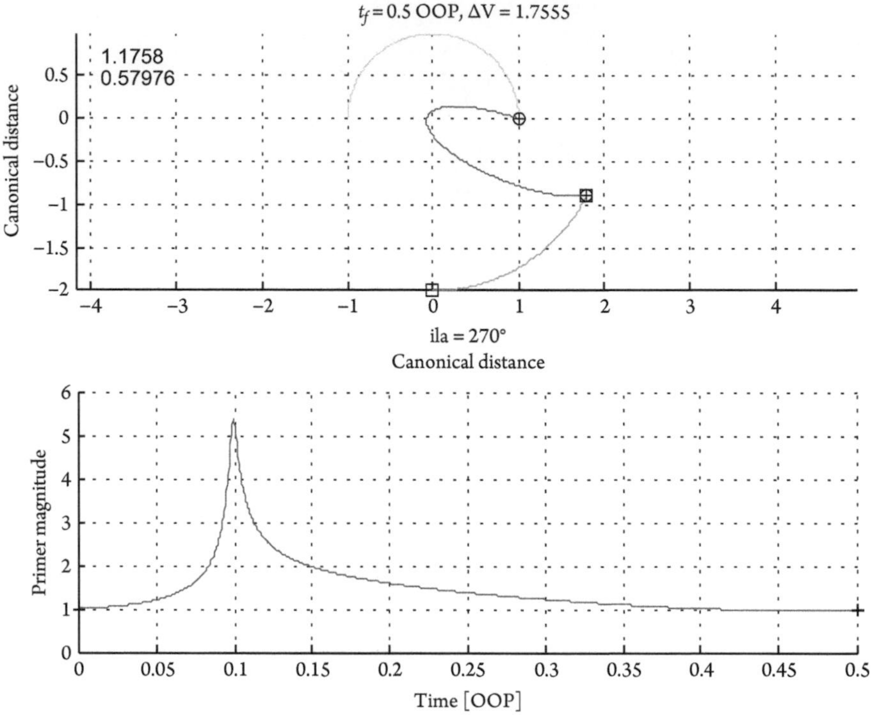

Figure 5.9 Indication of a need for a midcourse impulse

Figure 5.9 shows a primer-magnitude time history which indicates the need for a midcourse impulse.

The next step is to determine an initial position and time for the impulse. From Eq. (5.45) it is evident that for a given $\mathbf{p}_m$ the largest decrease in the cost is obtained by maximizing the value of the dot product, i.e., by choosing a position for the impulse that causes $\Delta \mathbf{v}_m$ to be parallel to the vector $\mathbf{p}_m$ and choosing the time t_m to be the time at which the primer magnitude has a maximum value. Choosing the position of the impulse so that the velocity change is in the direction of the primer vector sounds familiar because it is one of the necessary conditions derived previously, but how to determine this position is not at all obvious and we will have to derive an expression for this. Choosing the time t_m to be the time of maximum primer magnitude does not guarantee that the decrease in cost is maximized, because the value of Δv_m in the expression for dJ depends on the value of t_m. However, all we are doing is obtaining an initial position and time of the midcourse impulse to begin an iteration process. As long as our initial choice represents a decrease in the cost, we will opt for the simple device of choosing the time of maximum primer magnitude as our initial estimate of t_m.

Having determined t_m the position of the impulse, in terms of the $\delta \mathbf{r}_m$ to be added to $\mathbf{r}_m$ must also be determined. Obviously $\delta \mathbf{r}_m$ must be nonzero, otherwise the midcourse impulse would have zero magnitude. The property that must be satisfied in determining $\delta\mathbf{r}_m$ is that $\Delta\mathbf{v}_m$ be parallel to $\mathbf{p}_m$. To accomplish this, let us make use of the state transition matrix

$$\mathbf{\Phi}_{mo} \equiv \mathbf{\Phi}(t_m, t_o) = \begin{bmatrix} \mathbf{M}_{mo} & \mathbf{N}_{mo} \\ \mathbf{S}_{mo} & \mathbf{T}_{mo} \end{bmatrix} \tag{5.46}$$

to write

$$\delta\mathbf{v}_m^- = \mathbf{T}_{mo}\delta\mathbf{v}_o^+ \tag{5.47}$$

where $\delta\mathbf{r}_o = 0$ has been used, and analogously

$$\delta\mathbf{v}_m^+ = \mathbf{T}_{mf}\delta\mathbf{v}_f^- \tag{5.48}$$

In Eq. (5.48) it is evident that we need to use the $\mathbf{T}_{mf}$ partition of the transition matrix $\mathbf{\Phi}_{mf}$ which propagates the state from the final time back to the midcourse impulse time. It is the inverse of the matrix $\mathbf{\Phi}_{fm}$. While it is true that $\mathbf{\Phi}_{mf} = \mathbf{\Phi}_{fm}^{-1}$, in terms of the partitions, $\mathbf{T}_{mf} \neq \mathbf{T}_{fm}^{-1}$, i.e., a partition of the inverse matrix is not the inverse of the partition of the original matrix. However, because of the symplectic property of the transition matrix, we can use Eq. (4.32) to determine $\mathbf{T}_{mf} = \mathbf{M}_{fm}^T$. Then we can subtract Eq. (5.47) from Eq. (5.48) to yield an expression for the midcourse velocity change:

$$\begin{aligned} \Delta\mathbf{v}_m &= \delta\mathbf{v}_m^+ - \delta\mathbf{v}_m^- = \mathbf{T}_{mf}\delta\mathbf{v}_f^-\mathbf{T}_{mo}\delta\mathbf{v}_o^+ \\ &= \mathbf{M}_{fm}^T\delta\mathbf{v}_f^-\mathbf{T}_{mo}\delta\mathbf{v}_o^+ \end{aligned} \tag{5.49}$$

The variation in the midcourse impulse position can be represented by propagating the initial velocity variation forward to the impulse time and the final velocity variation backward to the impulse time:

$$\delta\mathbf{r}_m = \mathbf{N}_{mo}\delta\mathbf{v}_o^+ = \mathbf{N}_{mf}\delta\mathbf{v}_f^- \tag{5.50}$$

where the fact that $\delta\mathbf{r}_o = \delta\mathbf{r}_f = 0$ has been used. Solving for the velocity variations

$$\delta\mathbf{v}_o^+ = \mathbf{N}_{mo}^{-1}\delta\mathbf{r}_m \tag{5.51}$$

and

$$\delta\mathbf{v}_f^- = \mathbf{N}_{mf}^{-1}\delta\mathbf{r}_m = -\mathbf{N}_{fm}^{-T}\delta\mathbf{r}_m \tag{5.52}$$

where the symplectic property of the transition matrix has been used again. Combining Eqs. (5.49), (5.51), and (5.52)

$$\Delta\mathbf{v}_m = \mathbf{Q}\delta\mathbf{r}_m \tag{5.53}$$

where the matrix $\mathbf{Q}$ is defined as

$$\mathbf{Q} \equiv -\left(\mathbf{M}_{fm}^{T}\mathbf{N}_{fm}^{-T} + \mathbf{T}_{mo}\mathbf{N}_{mo}^{-1}\right) \tag{5.54}$$

In order to have $\Delta\mathbf{v}_m$ parallel to $\mathbf{p}_m$ it is necessary that $\Delta\mathbf{v}_m = \varepsilon\mathbf{p}_m$ with $\varepsilon > 0$. Combining this fact with Eq. (5.53) yields

$$\mathbf{Q}\delta\mathbf{r}_m = \Delta\mathbf{v}_m = \varepsilon\mathbf{p}_m \tag{5.55}$$

which yields the solution for $\delta\mathbf{r}_m$ as

$$\delta\mathbf{r}_m = \varepsilon\mathbf{Q}^{-1}\mathbf{p}_m \tag{5.56}$$

assuming $\mathbf{Q}$ is invertible.

The question that arises is how to select a value for ε. Obviously, too large a value will violate the linearity assumptions of the perturbation analysis. This is not addressed in Ref. [5.1], but we can maintain a small change by specifying

$$\frac{\delta r_m}{r_m} = \beta \tag{5.57}$$

where β is a specified small number such as 0.05. Eq. (5.55) then yields

$$\frac{\varepsilon\left|\mathbf{Q}^{-1}\mathbf{p}_m\right|}{r_m} = \beta \Longrightarrow \varepsilon = \frac{\beta r_m}{\left|\mathbf{Q}^{-1}\mathbf{p}_m\right|} \tag{5.58}$$

If the resulting $dJ \geq 0$, then decrease ε and try again. One should never accept a midcourse impulse position that does not decrease the cost, because a sufficiently small ε will always provide a lower cost.

The initial midcourse impulse position and time are now determined. We add the $\delta\mathbf{r}_m$ of Eq. (5.56) to the value of $\mathbf{r}_m$ on the nominal trajectory at the time t_m at which the primer magnitude has a maximum value that is greater than one. We next solve two Lambert Problems: From $\mathbf{r}_o$ to $\mathbf{r}_m + \delta\mathbf{r}_m$ and from $\mathbf{r}_m + \delta\mathbf{r}_m$ to $\mathbf{r}_f$.

The trajectory and primer history after the insertion of the initial midcourse impulse are shown in Fig. 5.10. Note that $\dot{\mathbf{p}}_m$ is discontinuous, which violates the NC.

The reason that $p_m = 1$ in Fig. 5.10 is that the final boundary condition of the first segment is $\mathbf{p}_{m_{new}} = \Delta\mathbf{v}_m/\Delta v_m = \mathbf{p}_{m_{old}}/p_{m_{old}}$ and this is also the initial boundary condition on the second segment.

5.1.3 Iteration on the midcourse impulse position and time

To determine how to efficiently iterate on the three components of position of the midcourse impulse and its time, one needs to derive expressions for the gradients of the cost

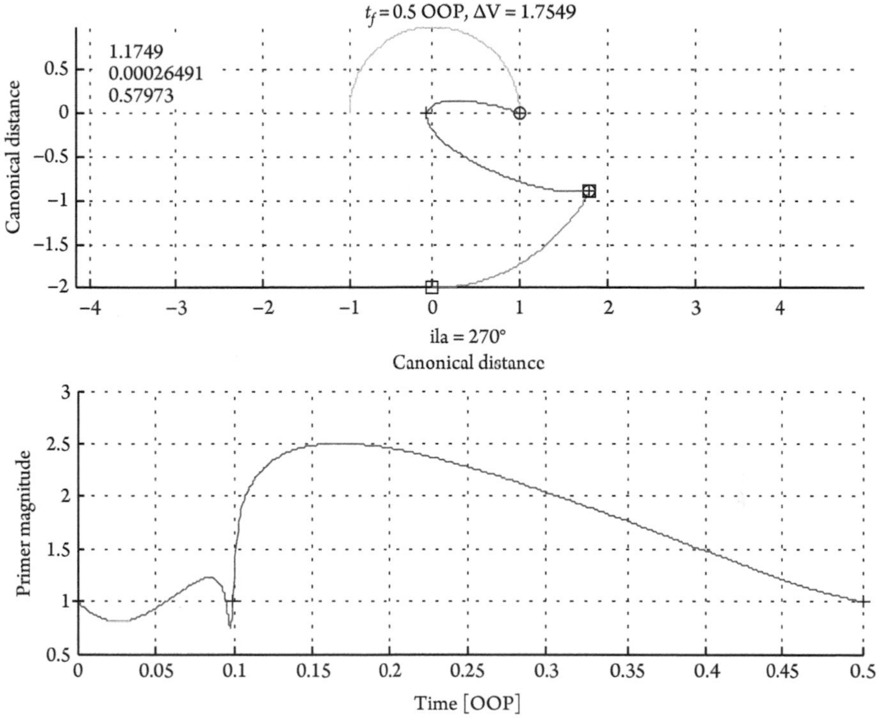

Figure 5.10 Initial (nonoptimal) three-impulse trajectory

with respect to these variables. To do this, one must compare the three-impulse trajectory (or three-impulse segment of an n-impulse trajectory) that resulted from the addition of he midcourse impulse with a perturbed three-impulse trajectory. Figure 5.11 shows the geometry, including the variations in the time dt_m and in the position $d\mathbf{r}_m$.

Note that, unlike a terminal coast, the values of $d\mathbf{r}_m$ and dt_m are *independent*. On an initial coast $d\mathbf{r}_o = \mathbf{v}_o^- dt_o$ and on a final coast $d\mathbf{r}_f = \mathbf{v}_f^+ dt_f$.

The cost on the nominal three-impulse trajectory is

$$J = \Delta v_o + \Delta v_m + \Delta v_f \tag{5.59}$$

and the variation in the cost due to perturbing the midcourse time and position is

$$dJ = \frac{\partial \Delta v_o}{\partial \Delta \mathbf{v}_o} d\Delta \mathbf{v}_o + \frac{\partial \Delta v_m}{\partial \Delta \mathbf{v}_m} d\Delta \mathbf{v}_m + \frac{\partial \Delta v_f}{\partial \Delta \mathbf{v}_f} d\Delta \mathbf{v}_f \tag{5.60}$$

which, as in Eq. (5.18), can be written as

$$dJ = \mathbf{p}_o^T d\Delta \mathbf{v}_o + \mathbf{p}_m^T d\Delta \mathbf{v}_m + \mathbf{p}_f^T d\Delta \mathbf{v}_f \tag{5.61}$$

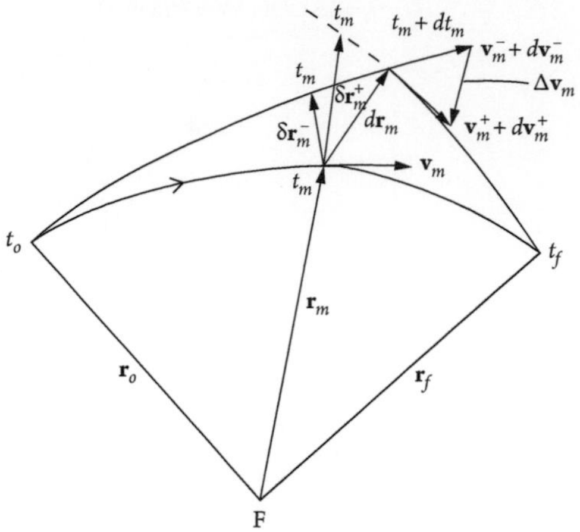

Figure 5.11 Iteration variables

Following the analysis from Eqs. (5.18) to (5.28) one notes that the middle term in Eq. (5.61) can be expressed as

$$d\Delta\mathbf{v}_m = d\mathbf{v}_m^+ - d\mathbf{v}_m^- \tag{5.62}$$

and that

$$d\mathbf{v}_m^\pm = \delta\mathbf{v}_m^\pm + \dot{\mathbf{v}}_m^\pm d\mathbf{t}_m \tag{5.63}$$

As in Eq. (5.22)

$$\dot{\mathbf{v}}_m^+ = \dot{\mathbf{v}}_m^- = \mathbf{g}(\mathbf{r}_m) \tag{5.64}$$

so that

$$d\Delta\mathbf{v}_m = \delta\mathbf{v}_m^+ - \delta\mathbf{v}_m^- \tag{5.65}$$

Next one uses Eq. (5.28) with $\delta\mathbf{r}_o = \delta\mathbf{r}_f = \mathbf{0}$ to yield

$$\mathbf{p}_o^T\delta\mathbf{v}_o^+ = \mathbf{p}_m^T\delta\mathbf{v}_m^{-1} - \dot{\mathbf{p}}_m^{T-}\delta\mathbf{r}_m^- \tag{5.66}$$

and

$$-\mathbf{p}_f^T\delta\mathbf{v}_f^- = \dot{\mathbf{p}}_m^{T+}\delta\mathbf{r}_m^+ - \mathbf{p}_m^T\delta\mathbf{v}_m^+ \tag{5.67}$$

where the fact that $\mathbf{p}_m$ is continuous at the midcourse impulse $(\mathbf{p}_m = \Delta\mathbf{v}_m/\Delta\mathrm{v}_m)$ has been used, but a discontinuity in $\dot{\mathbf{p}}_m$ has been allowed because there is no guarantee that it is continuous.

Substituting Eqs. (5.23, 5.24, 5.65, 5.66, 5.67) into Eq. (5.61) yields

$$dJ = \dot{\mathbf{p}}_m^{T+}\delta\mathbf{r}_m^+ - \dot{\mathbf{p}}_m^{T-}\delta\mathbf{r}_m^- \quad (5.68)$$

Finally, one can obtain a final form for the variation in cost by substituting

$$\delta\mathbf{r}_m^{\pm} = d\mathbf{r}_m - \mathbf{v}_m^{\pm}dt_m; \quad (5.69)$$

to obtain

$$dJ = (\dot{\mathbf{p}}_m^+ - \dot{\mathbf{p}}_m^-)^T d\mathbf{r}_m - (\dot{\mathbf{p}}_m^{T+}\mathbf{v}_m^+ - \dot{\mathbf{p}}_m^{T-}v_m^-)dt_m \quad (5.70)$$

which can be written more simply in terms of the Hamiltonian function $H = \dot{\mathbf{p}}^T\mathbf{v} - \mathbf{p}^T\mathbf{g}$ (for which the second term is continuous at t_m):

$$dJ = (\dot{\mathbf{p}}_m^+ - \dot{\mathbf{p}}_m^-)^T d\mathbf{r}_m - (H_m^+ - H_m^-)dt_m \quad (5.71)$$

Equation (5.71) provides the gradients of the cost with respect to the independent variations in position and time of the midcourse impulse for use in a nonlinear programming algorithm:

$$\frac{\partial J}{\partial \mathbf{r}_m} = (\dot{\mathbf{p}}_m^+ - \dot{\mathbf{p}}_m^-) \quad (5.72)$$

and

$$\frac{\partial J}{\partial t_m} = -(H_m^+ - H_m^-) \quad (5.73)$$

As a solution satisfying the NC is approached, the gradients tend to zero, in which case both the primer rate vector $\dot{\mathbf{p}}_m$ and the Hamiltonian function H_m become continuous at the midcourse impulse.

Note that when the NC are satisfied, the gradient with respect to t_m in Eq. (5.73) being zero implies that

$$H_m^+ - H_m^- = 0 = \dot{\mathbf{p}}_m^T(\mathbf{v}_m^+ - \mathbf{v}_m^-) = \dot{\mathbf{p}}_m^T\Delta\mathbf{v}_m = \Delta v_m\dot{\mathbf{p}}_m^T\mathbf{p}_m = 0 \quad (5.74)$$

which implies that $\dot{p}_m = 0$ indicating that the primer magnitude attains a local maximum value of unity. This is consistent with the NC that $p \leq 1$ and that $\dot{\mathbf{p}}$ be continuous.

Figure 5.12 shows the converged three-impulse optimal rendezvous that results from Fig. 5.10. Note that the final cost of 1.3681 is significantly less than the value of 1.7555 prior to adding the midcourse impulse. And the time of the midcourse impulse changed during the iteration from its initial value of 0.1 to a final value of approximately 0.17.

The absolute minimum cost time-open solution for the final radius and ila value of Fig. 5.12 is, of course, the Hohmann transfer as shown in Fig. 5.5. Its cost is significantly

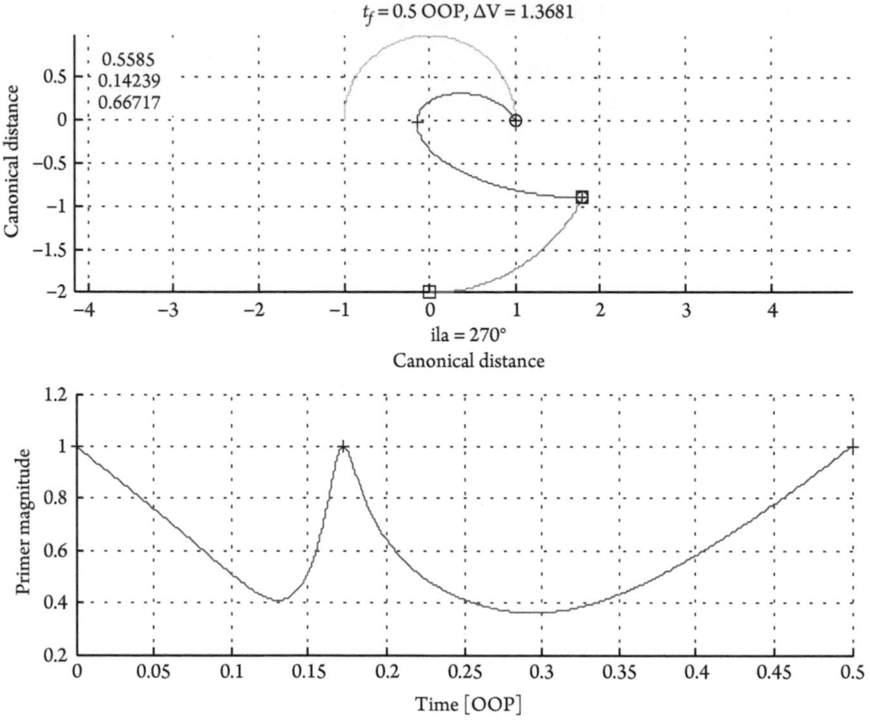

Figure 5.12 Optimal three-impulse trajectory and primer magnitude

less at 0.28446, but the transfer time is nearly three times as long at 1.8077 OOP. Of this 0.889 OOP is an initial coast required to achieve the correct target phase angle for the Hohmann transfer. Depending on the specific application, the total time required may be unacceptably long.

5.2 Fixed-time impulsive orbit transfer

In an orbit transfer there is no target body in contrast to a rendezvous or interception. The analysis that follows is a modification of the analysis in Section 5.1.

An important difference, however, is that $d\mathbf{r}_f \neq \mathbf{v}_f^+ dt_f$ due to the fact that there is no target body. In an orbit transfer $d\mathbf{r}_f$ and dt_f *are independent!*.

Starting with a two-impulse fixed-time solution

$$J = \Delta \mathrm{v}_o + \Delta \mathrm{v}_f \tag{5.75}$$

Eq. (5.18) becomes

$$dJ = \mathbf{p}_o^T \left(d\mathbf{v}_o^+ - d\mathbf{v}_o^-\right) + \mathbf{p}_f^T \left(d\mathbf{v}_f^+ - d\mathbf{v}_f^-\right) \tag{5.76}$$

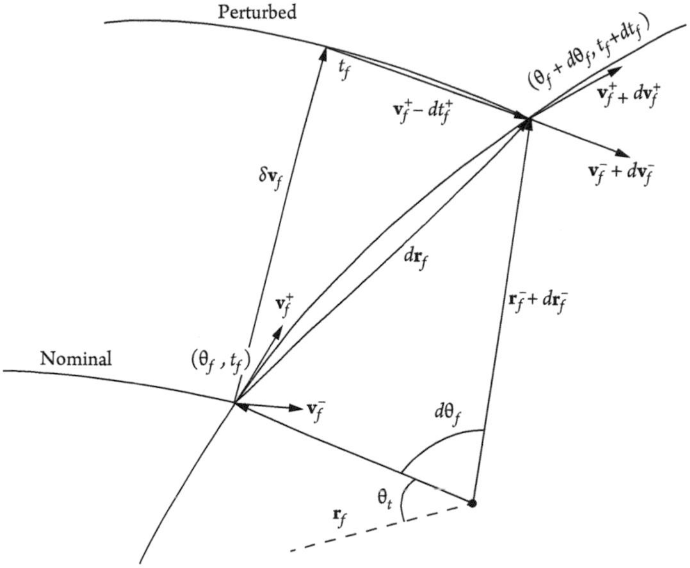

Figure 5.13 $d\mathbf{v}_f^+$ and $d\mathbf{r}_f$ determined by $d\theta_f$

but now $d\mathbf{v}_f^+ \neq \dot{\mathbf{v}}_f^+ dt_f = \mathbf{g}_f dt_f$ and $d\mathbf{r}_f \neq \mathbf{v}_f^+ dt_f$. Both $d\mathbf{v}_f^+$ and $d\mathbf{r}_f$ are determined by $d\theta_f$, (see Fig. 5.13). Similarly, $d\mathbf{v}_o^-$ and $d\mathbf{r}_o$ are determined by $d\theta_o$ on the initial orbit and $d\mathbf{v}_o^- \neq \dot{\mathbf{v}}_o dt_o = \mathbf{g}_o dt_o$ and $d\mathbf{r}_o \neq \mathbf{v}_o^- dt_o$.

As before,

$$d\mathbf{v}_f^- = \delta\mathbf{v}_f^- + \mathbf{g}_f dt_f \tag{5.77a}$$

and

$$d\mathbf{v}_o^+ = \delta\mathbf{v}_o^+ + \mathbf{g}_o dt_o \tag{5.77b}$$

so that Eq. (5.76) becomes

$$dJ = \mathbf{p}_f^T d\mathbf{v}_f^+ - \mathbf{p}_f^T \left(\delta\mathbf{v}_f^- + \mathbf{g}_f dt_f\right) + \mathbf{p}_o^T \left(\delta\mathbf{v}_o^+ + \mathbf{g}_o dt_o\right) - \mathbf{p}_o^T d\mathbf{v}_o^- \tag{5.78}$$

Equation (5.78) is in stark contrast to Eq. (5.25), which says that $dJ = \mathbf{p}_o^T \delta\mathbf{v}_o^+ - \mathbf{p}_f^T \delta\mathbf{v}_f^-$. As before, we use $\boldsymbol{\lambda}^T(t)\delta\mathbf{x}(t) = \text{constant} = \dot{\mathbf{p}}^T \delta\mathbf{r} - \mathbf{p}^T \delta\mathbf{v}$, so that

$$\mathbf{p}_o^T \delta\mathbf{v}_o^+ - \mathbf{p}_f^T \delta\mathbf{v}_f^- = \dot{\mathbf{p}}_o^T \delta\mathbf{r}_o - \dot{\mathbf{p}}_f^T \delta\mathbf{r}_f \tag{5.79}$$

Now Eq. (5.78) becomes

$$dJ = \mathbf{p}_f^T d\mathbf{v}_F^+ - \mathbf{p}_f^T \mathbf{g}_f dt_f + \mathbf{p}_o^T \mathbf{g}_o dt_o - \mathbf{p}_o^T d\mathbf{v}_o^- + \dot{\mathbf{p}}_o \delta\mathbf{r}_o - \dot{\mathbf{p}}_f^T \delta\mathbf{r}_f \tag{5.80}$$

Now, using $\delta \mathbf{r}_o = d\mathbf{r}_o - \mathbf{v}_o^+ dt_o$ and $\delta \mathbf{r}_f = d\mathbf{r}_f - \mathbf{v}_f^- dt_f$,

$$dJ = \dot{\mathbf{p}}_o^T d\mathbf{r}_o - \mathbf{p}_o^T d\mathbf{v}_o^- - H_o^+ dt_o - \left(\dot{\mathbf{p}}_f^T d\mathbf{r}_f - \mathbf{p}_f^T d\mathbf{v}_f^+ - H_f^- dt_f\right) \tag{5.81}$$

where the simplifications

$$H_o^+ = \dot{\mathbf{p}}_o^T \mathbf{v}_o^+ - \mathbf{p}_o^T \mathbf{g}_o \tag{5.82a}$$

and

$$H_f^- = \dot{\mathbf{p}}_f^T \mathbf{v}_f^- - \mathbf{p}_f^T \mathbf{g}_f \tag{5.82b}$$

have been used. The terms involving dt_o and dt_f show the usual interpretation

$$\frac{\partial J}{\partial t_o} = -H_o^+ \tag{5.83a}$$

and

$$\frac{\partial J}{\partial t_f} = H_f^- \tag{5.83b}$$

5.2.1 Circular terminal orbits

For a circular terminal orbit

$$d\mathbf{r} = \begin{bmatrix} dr \\ rd\theta \end{bmatrix} \quad \text{becomes} \quad \begin{bmatrix} 0 \\ rd\theta \end{bmatrix} \tag{5.84}$$

Defining the vector $d\boldsymbol{\theta}$ normal to the terminal orbit plane,

$$d\mathbf{r} = d\boldsymbol{\theta} \times \mathbf{r} \tag{5.85a}$$

and

$$d\mathbf{v} = d\boldsymbol{\theta} \times \mathbf{v} \tag{5.85b}$$

The vectors $\mathbf{r}$ and $\mathbf{v}$ are shown in Fig. 5.14. Note that $d\mathbf{r}$ is normal to $\mathbf{r}$ and $d\mathbf{v}$ is normal to $\mathbf{v}$.

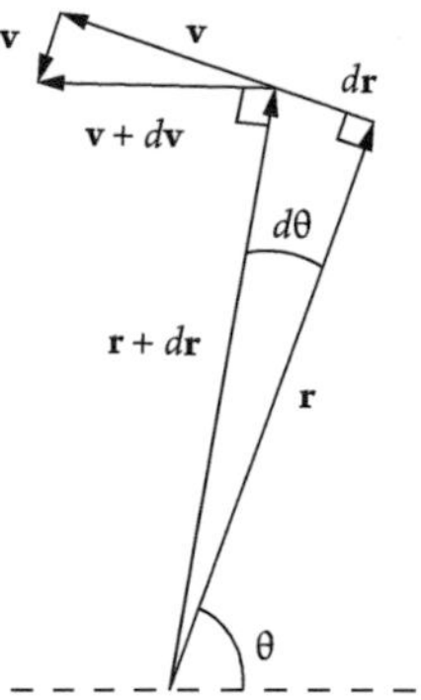

Figure 5.14 The effect of $d\theta$

Substituting into the terms in dJ:

$$-\dot{\mathbf{p}}_f^T d\mathbf{r}_f + \mathbf{p}_f^T d\mathbf{v}_f^+ = -\dot{\mathbf{p}}_f^T \left(d\boldsymbol{\theta}_f \times \mathbf{r}_f\right) + \mathbf{p}_f^T \left(d\boldsymbol{\theta}_f \times \mathbf{v}_f^+\right) \tag{5.86}$$

and use the identity $\mathbf{a}^T(\mathbf{b} \times \mathbf{c}) = \mathbf{b}^T(\mathbf{c} \times \mathbf{a})$ to write the terms in Eq. (5.86) as

$$-d\boldsymbol{\theta}_f^T \left(\mathbf{r}_f \times \dot{\mathbf{p}}_f\right) + d\boldsymbol{\theta}_f^T \left(\mathbf{v}_f^+ \times \mathbf{p}_f\right) \tag{5.87}$$

Define the vector $A \equiv \mathbf{p} \times \mathbf{v} - \dot{\mathbf{p}} \times \mathbf{r}$, as in Section 4.5, so that the terms in Eq. (5.87) become $-\mathbf{A}_f^{+^T} d\boldsymbol{\theta}_f$. The entire expression for dJ in Eq. (5.81) can now be written as

$$dJ = \mathbf{A}_o^{-^T} d\boldsymbol{\theta}_o - H_o^+ dt_o - \mathbf{A}_f^{+^T} d\boldsymbol{\theta}_f + H_f^- dt_f \tag{5.88}$$

which now provides an interpretation for $\mathbf{A}$ as

$$\frac{\partial J}{\partial \boldsymbol{\theta}_o} = \mathbf{A}_o^{-^T} \tag{5.89a}$$

and

$$\frac{\partial J}{\partial \boldsymbol{\theta}_f} = -\mathbf{A}_f^{+^T} \tag{5.89b}$$

More specifically, $\frac{\partial J}{\partial \theta_o}$ and $\frac{\partial J}{\partial \theta_f}$ (the gradients with respect to the scalars θ_o and θ_f) are given by the component of $\mathbf{A}_o^{-^T}$ normal to the initial orbit plane and the component of $-\mathbf{A}_f^{+^T}$ normal to the final orbit plane.

Equations (5.83a and b) and (5.89a and b) are the gradients of the cost J with respect to small changes in the terminal times and angles. They indicate what small changes in these variables will lower the cost.

A time-open optimal transfer will have the terminal values of the Hamiltonian equal to zero, and an angle-open optimal transfer will have the terminal magnitudes of the vector $\mathbf{A}$ equal to zero. An example of a time-open, angle-open optimal transfer is the Hohmann transfer (see Example 5.1).

Example 5.1 The Hohmann transfer

At an impulse for a Hohmann transfer the radius vector is orthogonal to the velocity vector. The primer vector is parallel to the velocity vector and the primer rate vector is parallel to the radius vector (sketch this). Using the definitions in Eqs. (5.82a) and (5.82b) and the interpretations in (5.83a) and (5.83b) $H_o^+ = H_f^- = 0$, indicating that the Hohmann transfer is a time-open optimal transfer. Also, using the definition of the vector $\mathbf{A}$ prior to Eq. (5.88) and the interpretations in (5.89a) and (5.89b) $\mathbf{A}_o = \mathbf{A}_f = 0$, indicating that the Hohmann transfer is also an angle-open optimal transfer.

(As a side-note, a simple proof of the global optimality of the Hohmann transfer using ordinary calculus rather than primer vector theory is given in Section B.2 of Appendix B.)

Problems

5.1 Verify that $\mathbf{T}_{mf} = \mathbf{M}_{fm}^T$, as used in Eq. (5.49) and also that $\mathbf{N}_{mf} = -\mathbf{N}_{fm}^T$.

5.2 For a *one-impulse intercept* ($J = \Delta v_o$) determine expressions for the gradients of the cost for an initial coast and for a variation in the intercept time. Determine an expression for dJ in terms of dt_o and dt_f.

5.3 For an inverse-square gravitational field:

a) Show that the quantity a derived in Ref. [5.1] is constant along an NT arc between impulses. The definition of a is $2\dot{\mathbf{p}}^T\mathbf{r} + \mathbf{p}^T\mathbf{v} - 3Ht$.

b) Demonstrate that the discontinuity in the variable a at an optimal interior or time- open terminal impulse is equal to the magnitude of the velocity change Δv.

c) Using the result of part (b) define a new variable by modifying a that is both constant between impulses and continuous at an optimal interior or time-open terminal impulse.

5.4 Consider an optimal interior or time-open terminal impulse in an inverse-square gravitational field. Determine a *numerical value* for the *maximum admissible angle* between the impulse direction and the local horizontal plane (the plane normal to the radius vector). Use the fact that the primer magnitude achieves a local maximum value of unity at the impulse. *Hint:* Determine an expression for the second derivative of the primer magnitude at the impulse point.

References

[5.1] Lion, P.M., and Handelsman, M., "Primer Vector on Fixed-Time Impulsive Trajectories", *AIAA Journal*, Vol. 6, No. 1, Jan 1968, pp. 127–32.

[5.2] Jezewski, D.J. and Rozendaal, H.L., An efficient method for calculating optimal free-space n-impulse trajectories. *AIAA Journal*, Vol 6, No. 11, Nov 1968, pp. 2160–5.

[5.3] Battin, R.H., *An Introduction to the Mathematics and Methods of Astrodynamics*, Revised Edition, AIAA Education Series, New York, 1999, Section 9.7.

[5.4] Prussing, J.E., and Conway, B.A., *Orbital Mechanics*, Oxford University Press, 1993, 2nd Edition 2012.

[5.5] Sandrik, S.L., *Primer-Optimized Results and Trends for Circular Phasing and Other Circle-to-Circle Impulsive Coplanar Rendezvous*, Ph.D. Thesis, University of Illinois at Urbana-Champaign, 2006.

6 Continuous-Thrust Trajectories

6.1 Quasi-circular orbit transfer

Several authors have presented simplified or approximate ways of determining continuous-thrust trajectories. In Ref. [6.1] Wiesel provides an approximate analysis of circle-to-circle coplanar orbit transfer using very low tangential continuous thrust. The approximation is that for constant thrust acceleration the transfer orbit is nearly circular, but with a very slowly changing radius.

The total orbital energy per unit mass in the two-body problem is given in Eq. (1.52) in Ref. [6.2] as

$$\varepsilon = -\frac{\mu}{2a} \tag{6.1}$$

Taking the time derivative

$$\dot{\varepsilon} = -\frac{\mu}{2a^2}\dot{a} \tag{6.2}$$

Due to a thrust acceleration $\mathbf{\Gamma}$, the rate of energy change can be written as

$$\dot{\varepsilon} = \mathbf{\Gamma} \cdot \mathbf{v} \tag{6.3}$$

(To see why this is true recall that the differential work done by a force is $dW = \mathbf{F} \cdot d\mathbf{r}$, and so the rate that work is done (the rate of energy change) is $\dot{W} = \mathbf{F} \cdot \mathbf{v}$). Because we are using energy per unit mass $\mathbf{\Gamma} = \mathbf{F}/m$ appears in Eq. (6.3).

Assuming a transfer from a lower to higher orbit, an increase in the orbital energy is required. For a constant mass flow rate, minimizing the transfer time will minimize the amount of propellant required. From Eq. (6.3) it can be seen that the way to maximize the *instantaneous* rate of increase of the energy is to align the thrust acceleration vector $\mathbf{\Gamma}$

Optimal Spacecraft Trajectories. John E. Prussing.

with the velocity vector **v**. (Actually, this does not yield the true minimum-time transfer between specified energy levels, especially for very low values of thrust acceleration.)

For a low thrust, the orbit remains nearly circular of instantaneous radius a (this is the approximation) and the velocity is given by

$$\mathrm{v} = \left(\frac{\mu}{a}\right)^{1/2} \tag{6.4}$$

With the thrust acceleration vector aligned with the velocity vector, Eqs. (6.2), (6.3) and (6.4) can be combined to yield:

$$\dot{\varepsilon} = \frac{\mu}{2a^2}\dot{a} = \Gamma\left(\frac{\mu}{a}\right)^{1/2} \tag{6.5}$$

or,

$$\dot{a} = 2\frac{\Gamma}{\mu^{1/2}}a^{3/2} \tag{6.6}$$

For a *constant* value of $\boldsymbol{\Gamma}$ one can separate variables and write

$$\int_{a_o}^{a} \xi^{-3/2}\, d\xi = \frac{2\Gamma}{\mu^{1/2}}\int_{t_o}^{t} d\eta \tag{6.7}$$

where ξ and η are dummy variables of integration.

Assuming $a > a_o$, integration yields

$$a_o^{-1/2} - a^{-1/2} = \frac{\Gamma}{\mu^{1/2}}(t - t_o) \tag{6.8}$$

So, letting $t_o = 0$, the time required to transfer between circular orbits of radii a_o and $a_f > a_o$ is given by

$$t_f = \frac{\mu^{1/2}}{\Gamma}(a_o^{-1/2} - a_f^{-1/2}) \tag{6.9}$$

The total $\Delta \mathrm{v}_{tot}$ required can be computed using the circular orbit condition of Eq. (6.4):

$$\mathrm{v}^2 = \frac{\mu}{a} \tag{6.10}$$

Differentiating and using Eq. (6.6),

$$2\mathrm{v}\dot{\mathrm{v}} = -\mu\frac{\dot{a}}{a^2} = -\frac{\mu}{a^2}\frac{2}{\mu^{1/2}}a^{3/2}\Gamma = -2\left(\frac{\mu}{a}\right)^{1/2}\Gamma$$

Using Eq. (6.4):

$$\dot{\mathrm{v}} = -\Gamma \tag{6.11}$$

that is, the spacecraft slows down linearly in time as local circular speed is maintained. So, the Δv_{tot} is calculated as

$$\Delta v_{tot} = v_o - v_f = \Gamma t_f = \left(\frac{\mu}{a_o}\right)^{1/2} - \left(\frac{\mu}{a_f}\right)^{1/2} \tag{6.12}$$

and is, interestingly, independent of the thrust acceleration Γ (but it still must be small in order for the approximations used to be valid), and is simply equal to the difference in the terminal circular orbit speeds.

Example 6.1 LEO to GEO transfer

Consider a continuous-thrust transfer from low earth orbit (LEO) to geosynchronous earth orbit (GEO). Assuming a 278-km altitude LEO, the terminal orbit radii are a_o = 6656 km and a_f = 42166 km. The gravitational constant μ for the earth is $3.986 \times 10^5\ km^3/s^2$. From Eq. (6.12) Δv_{tot} = 4.664 km/s compared to the (two-impulse) coplanar Hohmann transfer value Δv_H = 3.90 km/s. (The higher value for the low-thrust transfer is due to the gravity loss incurred by continuous thrusting.)

Assuming a realistic value for Γ of a "milligee" or 10^{-3} g (where 1 g = 9.80665 m/s^2) for a low-thrust engine, such as an ion engine, then Γ is about 10^{-2} m/s^2 or 10^{-5} km/s^2. From Eq. (6.12) one determines t_f to be 466393 s = 5.40 days. By contrast the Hohmann transfer time is the much smaller 5.27 hours.

So far, we have seen that the low-thrust transfer has a higher Δv_{tot} and a much longer transfer time than the Hohmann transfer. So why even consider using a low-thrust transfer? As we will see, the answer lies in the savings in required propellant (and the resultant increase in the payload mass delivered to the final orbit).

A realistic value for specific impulse (see Sect. 2.1) for an ion engine is 5000 s, which corresponds to an exhaust velocity c of about 50,000 m/s or 50 km/s. Recall that

$$\Gamma = bc/m \tag{6.13}$$

and our assumption that Γ is constant implies that the mass is assumed to be constant. (That's not really true, but we'll fix it later.) Let the value be m_o = 1000 kg.

From Eq. (6.13) we can determine the mass flow rate to be $b = 2 \times 10^{-4}$ kg/s. And, using the transfer time of 5.40 days computed above, the propellant consumed is $m_p = bt_f$ = 93.28 kg. For a Hohmann transfer using a high-thrust chemical engine a realistic value of specific impulse is I_{sp} = 300 s. Based on that value and the Δv_{tot} of 3.90 km/s, the mass of propellant required is 730.5 kg, *which is 637.22 kg more than the low-thrust requirement!* So, out of the 1000 kg in LEO *the continuous-thrust transfer can deliver 906.72 kg to GEO, compared with only 269.5 kg for the Hohmann transfer.* That's a dramatic increase of 236% that may well be worth the additional transfer time.

6.2 The effects of non-constant mass

Now, let's go back and include the effect of a non-constant mass on the continuous-thrust transfer. Both the propellant mass and the transfer time will be different. We can calculate these values easily, using known results from Eq. (6.9) of Ref. [6.2], the solution to the rocket equation (that accounts for the change in mass):

$$m_p = m_o(1 - e^{-\Delta v/c}) \tag{6.14}$$

We determine the propellant mass to be m_p = 89.06 kg, which is less than the value of 93.28 kg when a constant mass is assumed. This results in an additional 4.22 kg of final mass. That's almost 10 more pounds!

Based on Eq. (6.14) and the fact that $m_p = bt_f$ one determines that t_f = 5.15 days, about 6 hours less than the constant-mass value, due to the increased value of Γ.

In summary, for a LEO to GEO transfer the good news is that the final mass using the continuous-thrust transfer and accounting for the mass change is about 3.4 times that of the Hohmann transfer. The bad news is that the transfer time is much longer. But for cargo the additional final mass may be well worth it.

Continuous-thrust transfers have similar advantages for interplanetary applications.

6.3 Optimal quasi-circular orbit transfer

Consider the more complicated case of optimal (minimum-propellant) circle-to-inclined-circle transfer. Kechichian in Refs. [6.3] and [6.4] reformulated and extended an analysis of Edelbaum in Ref. [6.5] to determine an optimal transfer between non-coplanar circular orbits. Kechichian's reformulation is an optimal control problem that we have treated in Chapter 3. A low thrust is assumed and a quasi-circular approximation is made. A constant thrust acceleration **Γ** is assumed, so a minimum-time solution is also a minimum-propellant solution.

The Gaussian form of the Lagrange planetary equations (see Ref. [6.2]) for near-circular orbits is used with the thrust acceleration Γ being the small perturbing acceleration. By averaging over the fast variable (true anomaly) Edelbaum derives two equations:

$$\frac{di}{dt} = \frac{2\boldsymbol{\Gamma} \sin \beta}{\pi \mathrm{v}} \tag{6.15}$$

$$\frac{d\mathrm{v}}{dt} = -\boldsymbol{\Gamma} \cos \beta \tag{6.16}$$

where i is the orbital inclination, v is the scalar orbital velocity, and β is the (out-of-plane) thrust yaw angle. The value of β is assumed to be piecewise constant for each revolution switching sign at the orbital antinodes.

For the minimum-time cost functional we set $\phi = 0$ and $L = 1$ in Eq. (3.4). Then, using Eqs. (6.15–6.16) the Hamiltonian function in Eq. (3.7) is

$$H = 1 + 2\lambda_i \frac{\Gamma \sin\beta}{\pi \mathrm{v}} - \lambda_{\mathrm{v}} \Gamma \cos\beta \tag{6.17}$$

From Eq. (3.12) we have

$$\dot{\lambda}_i = -\frac{\partial H}{\partial i} = 0 \tag{6.18}$$

indicating that λ_i is constant.

Because the final time unspecified Eq. (3.47) applies with the result that $H_f = 0$. And because H is not an explicit function of time, Eq. (3.21) tells us that H is constant and equal to its final value of 0.

Using this result and the optimality condition Eq. (3.14) we can solve for λ_i and λ_{v}:

$$H = 0 = 1 + 2\lambda_i \frac{\Gamma \sin\beta}{\pi \mathrm{v}} - \lambda_{\mathrm{v}} \Gamma \cos\beta \tag{6.19a}$$

and

$$\frac{\partial H}{\partial \beta} = 0 = 2\lambda_i \frac{\Gamma \cos\beta}{\pi \mathrm{v}} + \lambda_{\mathrm{v}} \Gamma \sin\beta \tag{6.19b}$$

The result is

$$\lambda_i = -\frac{\pi \mathrm{v} \sin\beta}{2\Gamma} = constant \tag{6.20a}$$

and

$$\lambda_{\mathrm{v}} = \frac{\cos\beta}{\Gamma} \tag{6.20b}$$

Note that Eq. (6.20a) implies that v sin β is constant.

Using the interpretation given in Eq. (3.17) that the initial value $\lambda_k(t_o)$ is the sensitivity of the minimum cost to a small change in the initial state $x_k(t_o)$, Eqs. (20a and b) provide

$$\delta t_f = \lambda_{i_o} \delta i_o = -\frac{\pi \mathrm{v}_o \sin\beta_o}{2\Gamma} \delta i_o \tag{6.21}$$

and

$$\delta t_f = \lambda_{\mathrm{v}_o} \delta \mathrm{v}_o = \frac{\cos\beta_o}{\Gamma} \delta \mathrm{v}_o \tag{6.22}$$

The analysis above is continued in Refs. [6.3] and [6.4] to obtain a complete solution for a minimum-time solution to transfer from initial conditions (i_o, v_o) to specified final conditions (i_f, v_f), where v_o and v_f are the circular orbit velocities in the specified initial and final circular orbits, equal to $(\mu/a_o)^{1/2}$ and $(\mu/a_f)^{1/2}$.

The velocity on the minimum-time transfer orbit is determined to be:

$$v(t) = (v_o^2 - 2v_o\Gamma t\cos\beta_o + \Gamma^2 t^2)^{1/2} \tag{6.23}$$

where

$$\tan\beta_o = \frac{\sin[(\pi/2)\Delta i_f]}{(v_o/v_f) - \cos[(\pi/2)\Delta i_f]} \tag{6.24}$$

with $\Delta i_f = |i_o - i_f|$. Note that if $i_f > i_o$, $i(t) = i_o + \Delta i(t)$ and if $i_f < i_o$, $i(t) = i_o - \Delta i(t)$, where

$$\Delta i(t) = \frac{2}{\pi}\left[\tan^{-1}\left(\frac{\Gamma t - v_o\cos\beta_o}{v_o\sin\beta_o}\right) + \frac{\pi}{2} - \beta_o\right] \tag{6.25}$$

Also,

$$\tan\beta(t) = \frac{v_o\sin\beta_o}{v_o\cos\beta_o - \Gamma t} \tag{6.26}$$

and

$$\Delta v_{tot} = \left\{v_o^2 - 2v_f v_o\cos[(\pi/2)\Delta i_f] + v_f^2\right\}^{1/2} \tag{6.27}$$

Equation (6.27) is called *Edelbaum's equation.*

The fact that this solution actually minimizes the transfer time, rather than providing only a stationary value, is demonstrated in Ref. [6.6] using second-order conditions.

The computational algorithm is as follows: Given the gravitational constant μ, the terminal orbit radii a_o and a_f, the total inclination change desired Δi_f, and the low-thrust acceleration Γ, first determine the terminal circular orbit velocities v_o and v_f. Then determine Δv_{tot} from Eq. (6.27) and the transfer time $t_f = \Delta v_{tot}/\Gamma$.

To determine the time histories $v(t)$, $\Delta i(t)$, and $\beta(t)$ first determine β_o from Eq. (6.24) and then $v(t)$ from Eq. (6.23), $\Delta i(t)$ from Eq. (6.25), and $\beta(t)$ from Eq. (6.26).

Example 6.2 LEO to GEO transfer with plane change

The value of Earth's gravitational constant is $\mu = 398601.3$ km^3/s^2. The initial and final conditions are $a_0 = 7000$ km, $i_o = 28.5°$, $a_f = 42166$ km, $i_f = 0°$, and assume $\Gamma = 3.5 \times 10^{-7}$km/s^2. The transfer time is $t_f = 191.263$ days with corresponding $\Delta v_{tot} = 5.784$ km/s^2. The thrust yaw angle increases from its initial value $\beta_o = 21.98°$ to a final value 66.75°.

Problems

6.1 Use Eq. (6.9) to compute the time in days to escape the earth ($a_f \to \infty$) from the initial orbit in Example 6.1.

6.2 Specialize the results of Sec. 6.3 to the coplanar case $\Delta i_f = 0$. Determine the value of β_o and expressions for $\beta(t)$, $\mathrm{v}(t)$, and $\Delta \mathrm{v}_{tot}$. Compare these with expressions in Sec. 6.1.

6.3* Show that for arbitrary a_o, a_f, Γ, and $\Delta i_f = 1$ radian, $\Delta \mathrm{v}_{tot} = (\mathrm{v}_o^2 + \mathrm{v}_f^2)^{\frac{1}{2}}$, $\tan \beta_o = \mathrm{v}_f / \mathrm{v}_o$, and $\tan \beta_f = -\mathrm{v}_o / \mathrm{v}_f$.

6.4* Using $\mathrm{v}^2 = \mu / a$, derive an expression for $\dot{a}$ and show that if the specified plane change is sufficiently large, the value of a will exceed its final value and then decrease to its final value. Give a physical explanation for this.

6.5 In Ex. 6.2 give a physical reason for the final value of the thrust yaw angle being greater than its initial value.

6.6 Derive the numerical results cited in Ex. 6.2.

6.7 Perform an analysis analogous to Ex. 6.2 for the same terminal radii with $i_o = 90°$ and $i_f = 0°$.

6.8* For the transfer defined in Prob. 6.7 evaluate the time of maximum radius and calculate the value of the maximum radius. Comment on the implication of this maximum radius value.

6.9 Modify the parameters in Ex. 6.2 to be $a_o = 6656$ km, $i_o = 0$, and $\Gamma = 10^{-5}$km/s^2. Perform the calculations and compare the results with Ex. 6.1.

6.10 Demonstrate that the Minimum Principle discussed in Section 3.4 is satisfied by showing that $\partial^2 H / \partial \beta^2 > 0$.

6.11 Using Eq. (6.22), show that a small increase in a_o will result in a small decrease in t_f.

References

[6.1] Wiesel, W.E., *Spaceflight Dynamics*, 3rd Edition, Aphelion Press, 2010, pp. 97–9.

[6.2] Prussing, J.E., and Conway, B.A., *Orbital Mechanics*, 2nd Edition, Oxford Unive:rsity Press, 2012.

[6.3] Kechichian, J.A., "Reformulation of Edelbaum's Low-Thrust Transfer Problem Using Optimal Control Theory", *Journal of Guidance, Control, and Dynamics*, Vol. 20, No. 5, 1997, pp. 988–94.

[6.4] Kechichian, J.A., Chapter 14 "Optimal Low-Thrust Orbit Transfer" in *Orbital Mechanics*, 3rd Edition, V. Chobotov, Editor, AIAA Education Series, 2002.

[6.5] Edelbaum, T.N., "Propulsion Requirements for Controllable Satellites", *ARS Journal*, Aug. 1961, pp. 1079–89.

[6.6] Prussing, J.E., and Sandrik, S.L., "Second-Order Necessary Conditions and Sufficient Conditions Applied to Continuous-Thrust Trajectories", *Journal of Guidance, Control, and Dynamics*, Vol. 28, No. 4, 2005, pp. 812–16.

7 Cooperative Rendezvous

7.1 Continuous thrust cooperative rendezvous

Orbital rendezvous typically involves one thrusting (active) vehicle and one coasting (passive) target vehicle. However, a cooperative rendezvous, in which both vehicles provide thrust, can result in propellant savings if the masses and propulsive capabilities of the two vehicles are comparable.

Consider the cooperative rendezvous of two spacecraft (s/c). Assume that both s/c (#1 and #2) can provide thrust and they are "comparable" in mass and propulsive capability. (It would not make sense to try to maneuver the International Space Station to perform a rendezvous with a small s/c.)

In general, having both comparable s/c provide thrust will result in less total propellant expended for the rendezvous. This is because a rendezvous between an active and a passive (no thrust) s/c can be considered as a rendezvous of two active s/c where one is constrained to keep its thruster off. Allowing it to thrust removes that constraint, which in an optimization problem results in a lower (or equal) cost.

The terminal conditions for the cooperative rendezvous are that the final position vectors and velocity vectors of the two vehicles are equal:

$$\mathbf{r}_2(t_f) = \mathbf{r}_1(t_f); \quad \mathbf{v}_2(t_f) = \mathbf{v}_1(t_f) \tag{7.1}$$

where $\mathbf{r}_1(t_f)$ and $\mathbf{v}_1(t_f)$ are arbitrary. (One could consider the case where they are constrained in some way, but that is not done here.)

We will next show that for CSI and VSI propulsion systems the costs for cooperative rendezvous are *not* simply the sums of the costs of the individual s/c:

$$J_{CSI} \neq \int_{t_o}^{t_f} (\Gamma_1 + \Gamma_2)dt \tag{7.2}$$

Optimal Spacecraft Trajectories. John E. Prussing.

$$J_{VSI} \neq \frac{1}{2}\int_{t_o}^{t_f} (\Gamma_1^2 + \Gamma_2^2)dt \tag{7.3}$$

because, while "comparable" the two s/c are not "equal". For CSI $m_1 \approx m_2$ and $c_1 \approx c_2$ and for VSI $P_{1_{MAX}} \approx P_{2_{MAX}}$, where the m_i are the masses, the c_i are the exhaust velocities, and $P_{i_{MAX}}$ are the maximum available powers.

7.1.1 The CSI case

For the CSI case define

$$J(t) = m_1(t_o) - m_1(t) + m_2(t_o) - m_2(t) \tag{7.4}$$

and let

$$L = \dot{J} = -\dot{m}_1 - \dot{m}_2 = b_1 + b_2 \tag{7.5}$$

But $b_i = T_i/c_i = (m_i\Gamma_i)/c_i$ so

$$J_{CSI} = \int_{t_o}^{t_f} L\,dt = \int_{t_o}^{t_f} \left(\frac{m_1\Gamma_1}{c_1} + \frac{m_2\Gamma_2}{c_2}\right) dt \tag{7.6}$$

The ratios c_i/m_i are efficiencies of the two s/c. A higher value of c means there is more thrust T for a given mass flow rate b and a lower value of m means there is a larger acceleration for a given thrust T. And note that, whereas c is constant, m decreases as the s/c thrusts making the efficiencies time-varying and increasing.

The weighting coefficients m_i/c_i in the integrand of Eq. (7.6) are the reciprocals of the efficiencies causing the thrust acceleration for the less efficient s/c to get more weight in the cost (is penalized more). Next, we will apply the necessary conditions (NC) for an optimal cooperative CSI rendezvous with $0 \leq \Gamma_i \leq \Gamma_{i_{MAX}}$. The equations of motion are

$$\dot{\mathbf{r}}_1 = \mathbf{v}_1 \tag{7.7a}$$

$$\dot{\mathbf{r}}_2 = \mathbf{v}_2 \tag{7.7b}$$

$$\dot{\mathbf{v}}_1 = \mathbf{g}(\mathbf{r}_1) + \Gamma_1\mathbf{u}_1 \tag{7.7c}$$

$$\dot{\mathbf{v}}_2 = \mathbf{g}(\mathbf{r}_2) + \Gamma_2\mathbf{u}_2 \tag{7.7d}$$

where the $\mathbf{u}_i$ are unit vectors in the thrust directions.

Next form the Hamiltonian for the two-s/c system:

$$H = \sum_{i=1}^{2} \left(\frac{m_i}{c_i}\Gamma_i + \boldsymbol{\lambda}_{r_i}^T\mathbf{v}_i + \boldsymbol{\lambda}_{v_i}^T[\mathbf{g}(\mathbf{r}_i) + \Gamma_i\mathbf{u}_i]\right) \tag{7.8}$$

The terminal constraints are

$$\boldsymbol{\Psi} = \begin{bmatrix} \mathbf{r}_2(t_f) - \mathbf{r}_1(t_f) \\ \mathbf{v}_2(t_f) - \mathbf{v}_1(t_f) \end{bmatrix} = \begin{bmatrix} \boldsymbol{\Psi}_r \\ \boldsymbol{\Psi}_v \end{bmatrix} \tag{7.9}$$

Applying the NC

$$\dot{\boldsymbol{\lambda}}_{r_i}^T = -\frac{\partial H}{\partial \mathbf{r}_i} = -\boldsymbol{\lambda}_{v_i}{}^T \mathbf{G}(\mathbf{r}_i) \tag{7.10}$$

$$\dot{\boldsymbol{\lambda}}_{v_i}^T = -\frac{\partial H}{\partial \mathbf{v}_i} = -\boldsymbol{\lambda}_{r_i}{}^T \tag{7.11}$$

with boundary conditions given in terms of $\Phi = \boldsymbol{\nu}^T \boldsymbol{\Psi}$ as

$$\boldsymbol{\lambda}_{r_i}^T(t_f) = \frac{\partial \Phi}{\partial \mathbf{r}_i(t_f)} = \boldsymbol{\nu}^T \frac{\partial \boldsymbol{\Psi}}{\partial \mathbf{r}_i(t_f)} = \left[\boldsymbol{\nu}_r^T \boldsymbol{\nu}_v^T\right] \begin{bmatrix} \dfrac{\partial \boldsymbol{\Psi}_r}{\partial \mathbf{r}_i(t_f)} \\ \dfrac{\partial \boldsymbol{\Psi}_v}{\partial \mathbf{r}_i(t_f)} \end{bmatrix} \tag{7.12}$$

$$= \begin{cases} -\boldsymbol{\nu}_r^T, i = 1 \\ \boldsymbol{\nu}_r^T, \ i = 2 \end{cases} \tag{7.13}$$

Similarly,

$$\boldsymbol{\lambda}_{v_I}{}^T(t_f) = \begin{cases} -\boldsymbol{\nu}_v^T, & i = 1 \\ \boldsymbol{\nu}_v^T, & i = 2 \end{cases} \tag{7.14}$$

To minimize the Hamiltonian H of Eq. (7.8) choose the $\mathbf{u}_i$ opposite to the $\boldsymbol{\lambda}_{v_i}$ and define primer vectors $\mathbf{p}_i = -\boldsymbol{\lambda}_{v_i}$ for which $\dot{\mathbf{p}}_i = \boldsymbol{\lambda}_{r_i}$ as in the single s/c case. Then each s/c thrusts in the direction of its primer vector $\mathbf{u}_i = \mathbf{p}_i/p_i$ and its primer vector equation is $\ddot{\mathbf{p}}_i = \mathbf{G}(\mathbf{r}_i)\mathbf{p}_i$. According to Eq. (7.14) $\mathbf{u}_2(t_f) = -\mathbf{u}_1(t_f)$, as shown in Fig. 7.1.

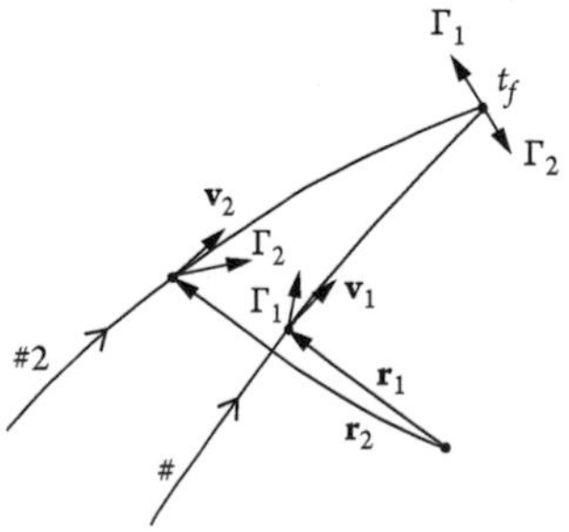

Figure 7.1 The optimal thrust directions at rendezvous

Using the optimal thrust directions the Hamiltonian becomes

$$\sum_{i=1}^{2}\left(\frac{m_i}{c_i}\Gamma_i + \dot{\mathbf{p}}_i^T\mathbf{v}_i - \mathbf{p}_i^T\mathbf{g}(\mathbf{r}_i) - p_i\Gamma_i\right) \tag{7.15}$$

The Hamiltonian is a linear function of the Γ_i so to minimize H over the choice of the Γ_i, define switching functions $S_i \equiv p_i - \frac{m_i}{c_i}$ and choose

$$\Gamma_i = \Gamma_{i_{MAX}} \quad \text{if} \quad S_i > 0 \tag{7.16a}$$

$$\Gamma_i = 0 \quad \text{if} \quad S_i < 0 \tag{7.16b}$$

For the more efficient s/c (larger c_i/m_i) the critical value m_i/c_i in the switching function is lower resulting in the more efficient s/c thrusting for longer periods of time.

Figure 7.2 shows a hypothetical switching function for each s/c and Fig. 7.3 shows switching functions for both s/c when s/c #1 is more efficient. Note that compared to s/c #2, s/c #1 has longer MT arcs and shorter NT arcs taking advantage of its greater efficiency.

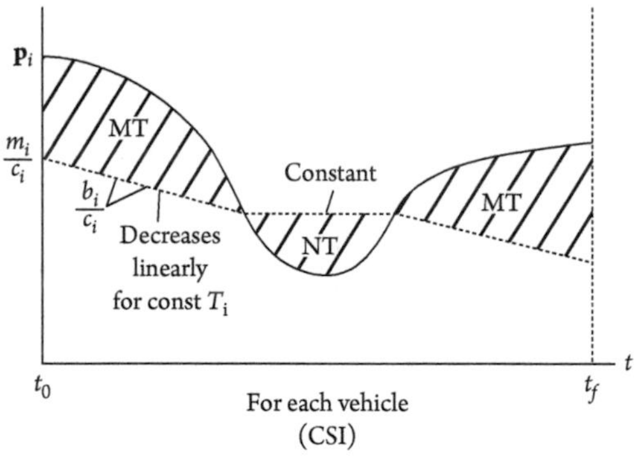

Figure 7.2 Single vehicle MT-NT-MT thrust sequence

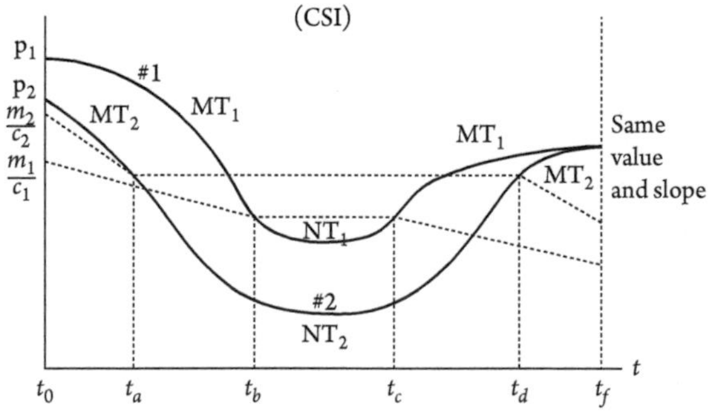

Figure 7.3 Two-vehicle MT-NT-MT thrust sequences

7.1.2 The VSI case

For the VSI case recall that

$$\frac{d}{dt}\left(\frac{1}{m}\right) = \frac{\Gamma^2}{2P} \tag{7.17}$$

so define

$$\dot{J} = L = \frac{d}{dt}\left(\frac{1}{m_1}\right) + \frac{d}{dt}\left(\frac{1}{m_2}\right) \tag{7.18}$$

Then

$$J = \frac{1}{m_{1_f}} + \frac{1}{m_{2_f}} - \frac{1}{m_{1_o}} - \frac{1}{m_{2_o}} \tag{7.19}$$

This cost has the same effect as maximizing the sum of the final masses because

$$\frac{\partial J}{\partial m_{i_f}} = -\frac{1}{{m_{i_f}}^2} < 0 \tag{7.20}$$

and

$$dJ = -\frac{1}{{m_{1_f}}^2} dm_{1_f} - \frac{1}{{m_{2_f}}^2} dm_{2_f} \tag{7.21}$$

So $dm_{i_f} > 0$ results in $dJ < 0$. Based on Eqs. (7.7) and (7.19) define the cost to be

$$J_{VSI} = \frac{1}{2}\int_{t_o}^{t_f}\left(\frac{\Gamma_1^2}{P_{1_{MAX}}} + \frac{\Gamma_2^2}{P_{2_{MAX}}}\right)dt \tag{7.22}$$

In J_{VSI} the $P_{i_{MAX}}$ are the efficiencies of the two s/c and the smaller value gets more weight in the cost.

To apply the NC for the VSI case, write the Hamiltonian as

$$H = \sum_{i-1}^{2}\left(\frac{1}{2}\frac{\Gamma_i^2}{P_{i_{MAX}}} + \boldsymbol{\lambda}_{r_i}{}^T\mathbf{v}_i + \boldsymbol{\lambda}_{v_i}{}^T[\mathbf{g}(\mathbf{r}_i) + \Gamma_i\mathbf{u}_i]\right) \tag{7.23}$$

The CSI NC Eqs. (7.9–14) also apply in the VSI case.

Using the optimal thrust directions the Hamiltonian of Eq. (7.23) becomes

$$H = \sum_{i=1}^{2}\left(\frac{1}{2}\frac{\Gamma_i^2}{P_{i_{MAX}}} + \dot{\mathbf{p}}_i^T\mathbf{v}_i - \mathbf{p}_i^T\mathbf{g}(\mathbf{r}_i) - p_i\Gamma_i\right) \tag{7.24}$$

To minimize H set

$$\frac{\partial H}{\partial \Gamma_i} = \frac{\Gamma_i}{P_{i_{MAX}}} - p_i = 0 \tag{7.25}$$

which, combined with $\mathbf{u}_i = \mathbf{p}_i/p_i$, results in the optimal thrust acceleration vector

$$\boldsymbol{\Gamma}_i(t) = P_{i_{MAX}}\mathbf{p}_i(t) \tag{7.26}$$

The larger efficiency $P_{i_{MAX}}$ results in a larger thrust acceleration. Note that the condition of Eq. (6.45) does minimize H because $\partial^2 H/\partial\Gamma_i^2 > 0$.

Note also that based on the discussion following Eq. (7.14), the two thrust acceleration vectors at the final time are oppositely directed.

7.2 Impulsive cooperative terminal maneuvers

The analysis here is a simplified and abbreviated version of that reported in Refs. [7.1] (expanded version with 11 figures) and [7.2] (original version with one figure). In those references, propellant mass fraction constraints are incorporated, but are not included here. These constraints make the problem more complicated, but also more interesting.

A thrust impulse provides an instantaneous velocity change $\Delta \mathrm{v}$. The relationship between propellant consumed Δm and the magnitude of the velocity change $\Delta \mathrm{v}$ is given in Ref. [7.3] as

$$\frac{\Delta m_i}{m_i} = 1 - e^{-\Delta \mathrm{v}_i/c_i} \tag{7.27}$$

where $1 = 1, 2$ represent the two s/c, the m_i are their masses prior to the terminal maneuver and the c_i are their exhaust velocities. Equation (7.27) indicates the well-known monotonic relationship between Δm and $\Delta \mathrm{v}$. Minimum total propellant consumption is achieved by minimizing $\Delta m_1 + \Delta m_2$.

Figure 7.4 shows the arrival velocities $\mathbf{v}_1$ and $\mathbf{v}_2$ of the two vehicles at the interception point, denoted by $\mathbf{r}_f$. These velocity vectors define points A and B. Point C represents a final velocity vector $\mathbf{v}_f$, achieved by the addition of vector velocity changes $\Delta\mathbf{v}_1$ and $\Delta\mathbf{v}_2$ to the arrival velocities $\mathbf{v}_1$ and $\mathbf{v}_2$, respectively. Because the final velocity vector $\mathbf{v}_f$ is unspecified the optimal location of point C must lie along the line segment AB connecting points A and B.

The reason for this can be seen in two ways. Consider a final velocity represented by a point not on the line segment AB, such as point D in Fig. 7.4. By projecting point D orthogonally onto the line segment AB at point C both $\Delta \mathrm{v}_i$ are made smaller indicating

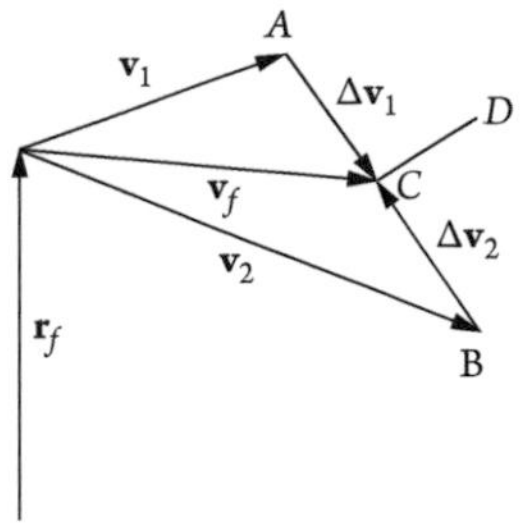

Figure 7.4 Velocity changes for an impulsive cooperative rendezvous

that both mass changes Δm_i are smaller, and therefore the total propellant consumed is smaller. Alternatively notice that for point C the total $\Delta \mathrm{v}$ is minimized and equal to

$$\Delta \mathrm{v}_1 + \Delta \mathrm{v}_2 = |\mathbf{v}_2 - \mathbf{v}_1| \equiv K \tag{7.28}$$

where K is defined as magnitude of the difference of the arrival velocities (equal to the length of the line segment AB).

Another way of stating this result is that the optimal final velocity $\mathbf{v}_f$ is a *convex combination* of the arrival velocities $\mathbf{v}_1$ and $\mathbf{v}_2$:

$$\mathbf{v}_f = \sigma \mathbf{v}_2 + (1-\sigma)\mathbf{v}_1 \quad 0 \leq \sigma \leq 1 \tag{7.29}$$

where $\sigma = \Delta \mathrm{v}_1/K$ and $1-\sigma = \Delta \mathrm{v}_2/K$.

For a given value of K the value of σ that minimizes total propellant consumption must be determined. The cost function $L = \Delta m_1 + \Delta m_2$ to be minimized can be written using Eq. (7.27) as

$$L(\sigma) = m_1(1 - e^{-\sigma K/c_1}) + m_2(1 - e^{-(1-\sigma)K/c_2}) \tag{7.30}$$

The two constraints are $\sigma \geq 0$ and $\sigma \leq 1$. In Section 1.3 we considered only a single inequality constraint, but here we have two. However, only one of the two can be active.

The cost L can be combined with the constraints as in Eq. (1.23) by defining

$$H(\sigma, \lambda_1, \lambda_2) = L(\sigma) - \lambda_1 \sigma + \lambda_2(\sigma - 1) \tag{7.31}$$

Then the necessary condition (1.24) is

$$\frac{\partial H}{\partial \sigma} = L'(\sigma) - \lambda_1 + \lambda_2 = 0 \tag{7.32}$$

with

$$\lambda_1 \geq 0, \quad i = 1, 2 \tag{7.33}$$

and

$$\lambda_1 = 0 \quad \text{if} \quad \sigma > 0 \tag{7.34a}$$

$$\lambda_2 = 0 \quad \text{if} \quad \sigma < 1 \tag{7.34b}$$

The first derivative term appearing in Eq. (7.32) is equal to

$$L'(\sigma) = \frac{Km_1}{c_1} e^{-\sigma K/c_1} - \frac{Km_2}{c_2} e^{-(1-\sigma)K/c_2} \tag{7.35}$$

The solutions to the necessary conditions Eqs. (7.32–34) are easily deduced using the property that the function $L(\sigma)$ is concave:

$$L''(\sigma) = -\frac{K^2 m_1}{c_1^2} e^{-\sigma K/c_1} - \frac{K^2 m_2}{c_2^2} e^{-(1-\sigma)K/c_2} < 0 \tag{7.36}$$

There are four cases of solutions to Eqs. (7.32–34):
Case I

$$\text{If}\, L'(0) \leq 0 \quad \text{then} \quad \sigma = 1 \tag{7.37a}$$

If $L'(0) \leq 0$ then $L'(1) < 0$ because of Eq. (7.36), $\lambda_1 = 0$, and $\lambda_2 = -L'(1) > 0$. Case I is shown in Fig. 7.5. In Figs. 7.5–7.7 $p_L = 0$ and $p_U = 1$.

Because of the concavity property, Case I represents a constrained global maximum of L at $\sigma = p_L$ and the desired constrained global minimum at $\sigma = p_U$ as shown in Fig. 3. The necessary conditions (7.33 and 7.34) are satisfied by

$$\lambda_1 = 0 \quad \text{and} \quad \lambda_2 = -L'(p_L) \geq 0 \tag{17a}$$

Case II

$$\text{If}\, L'(1) \geq 0 \quad \text{then} \quad \sigma = 0 \tag{7.37b}$$

If $L'(1) \geq 0$ then $L'(0) > 0$, $\lambda_2 = 0$, and $\lambda_1 = L'(0) > 0$. Case II is shown in Fig. 7.6.

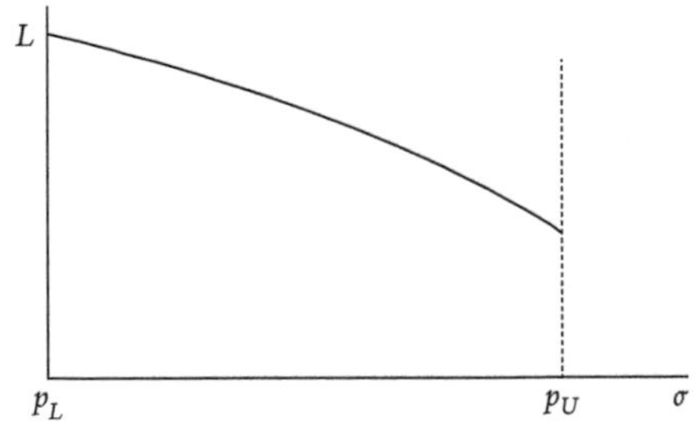

Figure 7.5 Case I

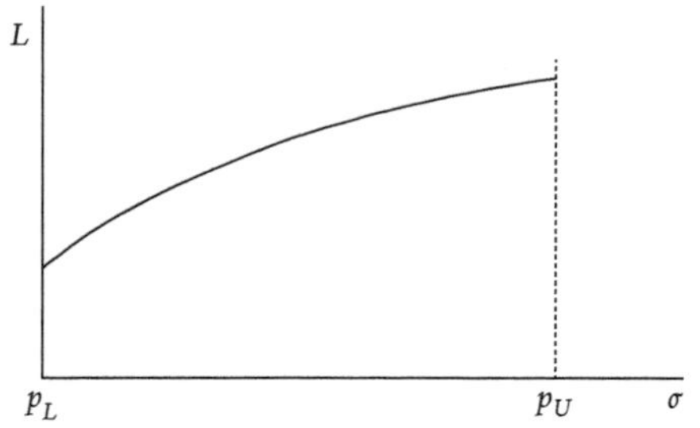

Figure 7.6 Case II

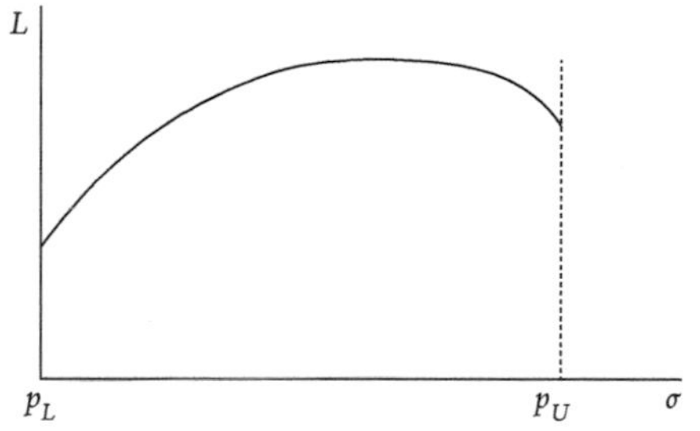

Figure 7.7 Cases III and IV

Case II represents a constrained global maximum at $\sigma = p_U$ and the desired global minimum at $\sigma = p_L$ as shown in Fig. 4. The necessary conditions (7.33 and 7.34) are satisfied by

$$\lambda_2 = 0 \quad \text{and} \quad \lambda_1 = L'(p_U) \geq 0 \tag{7.37b}$$

Case III

$$\text{If} \quad L'(0) > 0 \quad \text{and} \quad L'(1) < 0 \quad \text{then} \quad \sigma = \begin{Bmatrix} 0 \text{ if } L(0) < L(1) \\ 1 \text{ if } L(0) > L(1) \end{Bmatrix} \tag{7.37c}$$

In the $\sigma = 0$ solution $\lambda_2 = 0$, $\lambda_1 = L'(0) > 0$. In the $\sigma = 1$ solution $\lambda_1 = 0$, $\lambda_2 = -L'(1) > 0$. Both solutions are local constrained minima.

Case IV

Another solution to Eq. (7.32) is $\lambda_1 = \lambda_2 = L'(\sigma) = 0$. This is the unconstrained stationary cost, but because of Eq. (7.36) it is a maximum. Both Cases III and IV are shown in Fig. 7.7.

The maximum represented by $\lambda_1 = \lambda_2 = 0$ (both constraints inactive) and $L' = 0$ illustrates the hazards of using only first-order necessary conditions for a minimization.

Both local minima also satisfy the first-order necessary conditions. The desired minimizing solution is either $\sigma = p_L$ (for which $\lambda_2 = 0$ and $\lambda_1 = L'(p_L) \geq 0$ or $\sigma = p_U$ (for which $\lambda_1 = 0$ and $\lambda_2 = -L'(p_U) \geq 0$, whichever provides the lower cost.

In the minimization Cases I, II, and III, all of the velocity change is provided by a single s/c and $\lambda_i > 0$ for the active constraints satisfying the sufficient condition mentioned after Eq. (1.28).

As discussed in section 7.1.1 the efficiency of each s/c is the ratio c_i/m_i. A higher exhaust velocity and a smaller mass result in a larger velocity change for a given amount of mass expelled. This is the reason that only one s/c provides all the required velocity change. The exhaust velocity c is constant and the mass m decreases during thrusting increasing the s/c efficiency. That also explains the maximum propellant Case IV. When both s/c thrust each one's efficiency is increased only partially.

Problems

7.1 Fig. 7.4 consistent with Eq. (7.14) and Fig. 7.1? Explain your answer.

7.2 Consider a cooperative impulsive rendezvous in the $x - y$ plane. The two vehicles meet at the origin with velocities

$$\mathbf{v}_1 = [11]$$

and

$$\mathbf{v}_2 = [1 - 1]$$

Determine:

(a) The optimal value of $\Delta m_{tot} = \Delta m_1 + \Delta m_2$.

(b) The optimal $\mathbf{v}_f$ after the rendezvous.

(c) The value of Δm_{tot} if vehicle 1 does the entire maneuver.

(d) The value of Δm_{tot} if vehicle 2 does the entire maneuver.

(e) The value of Δm_{tot} if $\Delta \mathrm{v}_2 = 2\,\Delta \mathrm{v}_1$.

(f) The stationary value of Δm_{tot}, if any.

For three cases:

(1) $m_1 = 2/3, m_2 = 1/3; c_1 = c_2 = 1$

(2) $m_1 = 2/3, m_2 = 1/3; c_1 = 1, c_2 = 1/3$

(3) $m_1 = m_2 = 1/2; c_1 = c_2 = 1$

References

[7.1] Prussing, J.E., "Terminal Maneuver for an Optimal Cooperative Impulsive Rendezvous", AAS Paper 97-649, AAS/AIAA Astrodynamics Specialist Conference, Sun Valley, ID, August 1997, available in Volume 97 *Advances in the Astronautical Sciences*, pp. 759–74, Univelt Publishing Inc., San Diego.

[7.2] Prussing, J.E. and Conway, B.A., "Optimal Terminal Maneuver for a Cooperative Impulsive Rendezvous", *Journal of Guidance, Control, and Dynamics*, Vol. 12, No. 3, May–Jun 1989, pp. 433–5, also *Errata*, Vol. 12, No. 4, Jul–Aug 1989, p. 608.

[7.3] Prussing, J.E. and Conway, B.A., *Orbital Mechanics*, Oxford University Press, 2nd Edition, 2012, Chapter 6.

8 Second-Order Conditions

8.1 Second-order NC and SC for a parameter optimization

Let's review the material in Section 1.1 using a simple scalar function $f(x)$. Assume $f(x)$ is continuous and differentiable ("smooth") and there are no constraints on x. In order to minimize the value of $f(x)$ the first-order NC is

$$f'(x) = 0 \tag{8.1}$$

and this condition provides one or more stationary values x^*.

A second-order NC is

$$f''(x^*) \geq 0 \tag{8.2}$$

and a second-order SC is

$$f''(x^*) > 0 \tag{8.3}$$

Why is Eq. (8.2) a NC? Because if it is not satisfied the function has a maximum value at x^*. So to be a minimum it is necessary that it not be a maximum!

If we fail to satisfy a NC, the solution is definitely not a minimum. But if we fail to satisfy a SC, the result is inconclusive and further testing is required, as illustrated in Example 8.1.

Example 8.1 Simple scalar example

Consider $f(x) = x^4$. The NC of Eq. (8.1) yields $f'(x) = 4x^3 = 0$, for which there are three equal solutions $x^* = 0$. The NC of Eq. (8.2) is satisfied because $f''(x^*) = 12x^{*3} = 0$, but the SC of Eq. (8.3) is not, because $f''(x^*)$ is not > 0. Does $x^* = 0$ minimize $f(x) = x^4$? We don't know and further testing is required.

Optimal Spacecraft Trajectories. John E. Prussing.

This requires a Taylor series expansion about x^*:

$$f(x) = f(x^*) + f'(x^*)(x - x^*) + \frac{1}{2!}f''(x^*)(x - x^*)^2 + \frac{1}{3!}f'''(x^*)(x - x^*)^3 + \frac{1}{4!}f^{iv}(x^*)(x - x^*)^4 + \cdots \tag{8.4}$$

Because $f'''(x^*) = 24x^* = 0$ only the fourth derivative term is nonzero because $f^{iv}(x^*) = 24$. So for any $x \neq x^* f(x) > f(x^*)$ in Eq. (8.4) and $x^* = 0$ provides a minimum value, as verified by a simple sketch of the function. This also points out an incorrect statement in some elementary calculus books that "if $f''(x^*) = 0 x^*$ is an inflection point". (See Problem 8.1 for $f(x) = x^3$.)

8.2 The second variation in an optimal control problem

For an optimal control problem we will examine the *second variation* of the cost J: $\delta(\delta J) = \delta^2 J$. Recall, as discussed in Section 3.2 the first variation $\delta J = 0$ due to $\delta \mathbf{u}(t)$ and subject to constraints provides first-order NC for a minimum of J. The solution provided by $\delta J = 0$ is a stationary solution.

We will analyse the *accessory minimum* problem: minimize $\delta^2 J$ subject to linearized equations of motion $\dot{\mathbf{x}} = \mathbf{f}(\mathbf{x}, \mathbf{u}, t)$ and constraints $\boldsymbol{\psi}[\mathbf{x}(t_f), t_f] = \mathbf{0}$:

$$\delta \dot{\mathbf{x}} = \mathbf{f}_x \delta \mathbf{x} + \mathbf{f}_u \delta \mathbf{u} \tag{8.5}$$

$$\delta \boldsymbol{\psi} = \mathbf{0} \tag{8.6}$$

Conditions for which $\delta^2 J \geq 0$ subject to the constraints with minimum value of 0 only for $\delta \mathbf{u} = 0$ are part of the SC for a minimum of J.

For this problem the cost $\delta^2 J$ is quadratic in $\delta \mathbf{x}$ and $\delta \mathbf{u}$, and the constraint equation of motion is linear in $\delta \mathbf{x}$ and $\delta \mathbf{u}$, which is the same form as the linear-quadratic problem (LQP) in optimal control. So we will follow the same formalism to solve the problem.

In this analysis the notation will be slightly different than in Chapter 3 for the terminal constraint. In this chapter the constraint $\boldsymbol{\psi}[\mathbf{x}(t_f), t_f] = \mathbf{0}$ is a $(q + 1) \times 1$ vector, where $q = 0$ for a single terminal constraint and $q > 0$ for multiple constraints.

For $q = 0$ either t_f is explicitly specified or the single terminal constraint acts as a "stopping condition" that implicitly specifies t_f, e.g., achieving a specified final altitude. So for $q = 0$, $\delta \mathbf{u}(t)$ can be treated as arbitrary (unconstrained) and there is always at least one terminal constraint.

Forming the second variation follows our Section 3.2 and Bryson and Ho (Ref. [8.2]). Form the augmented cost

$$\bar{J} = \Phi + \int [L + \boldsymbol{\lambda}^T(\mathbf{f} - \dot{\mathbf{x}})]dt \tag{8.7}$$

where $\Phi = \phi + \boldsymbol{\nu}^T \boldsymbol{\psi}$.

Using $H = L + \boldsymbol{\lambda}^T\mathbf{f}$ we want to minimize the second variation

$$\delta^2 J = \frac{1}{2}\delta\mathbf{x}^T(t_f)\Phi_{x_f x_f}\delta\mathbf{x}(t_f) + +\frac{1}{2}\int [\delta\mathbf{x}^T\delta\mathbf{u}^T]\begin{bmatrix} H_{xx} & H_{xu} \\ H_{ux} & H_{uu}\end{bmatrix}\begin{bmatrix}\delta\mathbf{x} \\ \delta\mathbf{u}\end{bmatrix} \quad (8.8)$$

where $\Phi_{x_f x_f}$ is a symmetric matrix having elements $\partial^2\Phi/\partial x_{f_i} x_{f_j}$.

Subject to the variational equation of $\dot{\mathbf{x}} = \mathbf{f}$:

$$\delta\dot{\mathbf{x}} = \mathbf{f}_x\delta\mathbf{x} + \mathbf{f}_u\delta\mathbf{u}; \quad \delta\mathbf{x}(t_o) = \mathbf{0} \quad (8.9a)$$

and the variational equation of $\dot{\boldsymbol{\lambda}}^T = -H_x$:

$$\delta\dot{\boldsymbol{\lambda}} = -H_{xx}\delta\mathbf{x} - \mathbf{f}_x^T\delta\boldsymbol{\lambda} - H_{xu}\delta\mathbf{u}$$

with

$$\delta\boldsymbol{\lambda}(t_f) = \Phi_{x_f x_f}\delta\mathbf{x}(t_f) + \boldsymbol{\psi}_{x_f} d\mathbf{v} \quad (8.9b)$$

and the variation of $H_{\mathbf{u}} = \mathbf{0}^T$:

$$H_{ux}\delta\mathbf{x} + \mathbf{f}_u^T\delta\boldsymbol{\lambda} + H_{uu}\delta\mathbf{u} = \mathbf{0} \quad (8.9c)$$

and the variation of ψ:

$$\delta\boldsymbol{\psi} = \psi_{x_f}\delta\mathbf{x}(t_f) \quad (8.9d)$$

So now we solve Eq. (8.9c)

$$\delta\mathbf{u} = -H_{uu}^{-1}(H_{ux}\delta\mathbf{x} + \mathbf{f}_u^T\boldsymbol{\lambda}) \quad (8.10)$$

where H_{uu} must be nonsingular. Substitute $\delta\mathbf{u}$ into Eq. (8.9a)

$$\begin{aligned}\delta\dot{\mathbf{x}} &= \mathbf{f}_x\delta\mathbf{x} - \mathbf{f}_u H_{uu}^{-1}(H_{ux}\delta\mathbf{x} + \mathbf{f}_u^T\boldsymbol{\lambda}) \\ &= A_1\delta\mathbf{x} - A_2\delta\boldsymbol{\lambda}\end{aligned} \quad (8.11a)$$

where

$$A_1 \equiv \mathbf{f}_x - \mathbf{f}_u H_{uu}^{-1} H_{ux} \quad (8.11b)$$

and

$$A_2 \equiv \mathbf{f}_u H_{uu}^{-1}\mathbf{f}_u^T \quad (8.11c)$$

Substituting $\delta\mathbf{u}$ in Eq. (8.9b)

$$\begin{aligned}\delta\dot{\boldsymbol{\lambda}} &= -H_{xx}\delta\mathbf{x} - \mathbf{f}_u^T\delta\boldsymbol{\lambda} + H_{xu}H_{uu}^{-1}(H_{ux}\delta\mathbf{x} + \mathbf{f}_u^T\delta\boldsymbol{\lambda}) \\ &= -A_o\delta\mathbf{x} - A_1^T\delta\boldsymbol{\lambda}\end{aligned} \tag{8.12a}$$

where

$$A_o \equiv H_{xx} - H_{xu}H_{uu}^{-1}H_{ux} \tag{8.12b}$$

and H_{xu} is the $n \times m$ matrix $\partial(H_x^T)/\partial\mathbf{u}$ and $H_{ux}^T = H_{xu}$.

Combining Eqs. (8.11a) and (8.12)

$$\begin{bmatrix}\delta\dot{\mathbf{x}} \\ \delta\dot{\boldsymbol{\lambda}}\end{bmatrix} = \begin{bmatrix}A_1 & -A_2 \\ -A_o & -A_1^T\end{bmatrix}\begin{bmatrix}\delta\mathbf{x} \\ \delta\boldsymbol{\lambda}\end{bmatrix} \equiv P(t)\begin{bmatrix}\delta\mathbf{x} \\ \delta\boldsymbol{\lambda}\end{bmatrix} \tag{8.13}$$

where $P(t)$ is a $2n \times 2n$ matrix. The solution to Eq. (8.13) can be written in terms of a transition matrix as

$$\begin{bmatrix}\delta\mathbf{x}(t) \\ \delta\boldsymbol{\lambda}(t)\end{bmatrix} = \Theta(t,\tau)\begin{bmatrix}\delta\mathbf{x}(\tau) \\ \delta\boldsymbol{\lambda}(\tau)\end{bmatrix} \tag{8.14a}$$

where, as in Eq. (D.11) of Appendix D

$$\dot{\Theta}(t,\tau) = P(t)\Theta(t,\tau) \tag{8.14b}$$

and

$$\Theta(\tau,\tau) = \mathbf{I}_{2n} \tag{8.14c}$$

Note that from Eq. (8.12b)

$$\begin{aligned}A_o^T &= H_{xx}^T - H_{ux}^T H_{uu}^{-T} H_{xu}^T \\ &= H_{xx} - H_{xu}H_{uu}^{-1}H_{ux} = A_o\end{aligned} \tag{8.15a}$$

and therefore A_o is symmetric.

Also, from Eq. (8.11c)

$$A_2^T = \mathbf{f}_u H_{uu}^{-T}\mathbf{f}_u^T = \mathbf{f}_u H_{uu}^{-1}\mathbf{f}_u^T = A_2 \tag{8.15b}$$

and therefore A_2 is symmetric.

These symmetry properties and the form of the matrix $P(t)$ in Eq. (8.13) imply that the transition matrix $\Theta(t,\tau)$ in Eq. (8.14) is symplectic (see Problem 4.9).

8.3 Review of the linear-quadratic problem

Here we will review a simple case of the LQP, namely a fixed final time t_f and no terminal constraints of the form $\boldsymbol{\psi} = \mathbf{0}$. The terms in the (quadratic) cost are

$$\phi = \frac{1}{2}\mathbf{x}^T(t_f)S_F\mathbf{x}(t_f) \tag{8.16a}$$

where the matrix S_F is $n \times n$ and

$$L = \frac{1}{2}\mathbf{x}^T A\mathbf{x} + \frac{1}{2}\mathbf{u}^T B\mathbf{u} + \frac{1}{2}\mathbf{x}^T N\mathbf{u} + \frac{1}{2}\mathbf{u}^T N^T\mathbf{x} \tag{8.16b}$$

Note that the last two terms in L involving the $n \times m$ matrix N are not in the usual LQP, but we need them for the analysis of the second variation.

The state equation is

$$\dot{\mathbf{x}} = F\mathbf{x} + G\mathbf{u} \tag{8.17}$$

As discussed in Appendix F the matrices S_F, A, and B that appear in quadratic forms can be assumed to be symmetric.

To apply the first-order NC we form

$$H = \frac{1}{2}\mathbf{x}^T A\mathbf{x} + \frac{1}{2}\mathbf{u}^T B\mathbf{u} + \frac{1}{2}\mathbf{x}^T N\mathbf{u} + \frac{1}{2}\mathbf{u}^T N^T\mathbf{x} + \boldsymbol{\lambda}^T(F\mathbf{x} + G\mathbf{u}) \tag{8.18}$$

Applying the NC

$$\frac{\partial H}{\partial \mathbf{u}} = \mathbf{u}^T B + \mathbf{x}^T N + \boldsymbol{\lambda}^T G = \mathbf{0}^T \tag{8.19a}$$

Transposing

$$B\mathbf{u} + N^T\mathbf{x} + G^T\boldsymbol{\lambda} = \mathbf{0} \tag{8.19b}$$

where we have assumed that the matrix B is symmetric. Next

$$\dot{\boldsymbol{\lambda}}^T = -\frac{\partial H}{\partial \mathbf{x}} = -\mathbf{x}^T A - \mathbf{u}^T N^T - \boldsymbol{\lambda}^T F \tag{8.20a}$$

Transposing

$$\dot{\boldsymbol{\lambda}} = -A\mathbf{x} - N\mathbf{u} - F^T\boldsymbol{\lambda} \tag{8.20b}$$

with

$$\boldsymbol{\lambda}^T(t_f) = \frac{\partial \phi}{\partial \mathbf{x}(t_f)} = \mathbf{x}^T(t_f)S_F \tag{8.20c}$$

and, finally

$$\dot{\mathbf{x}} = F\mathbf{x} + G\mathbf{u}; \quad \mathbf{x}(t_o) = \mathbf{x}_o \tag{8.21}$$

Once again we have a two-point boundary value problem (2PBVP) (as in Section 3.2) because the boundary conditions are split between the initial and final times.

To solve this LQP first solve Eq. (8.19b) for

$$\mathbf{u} = -B^{-1}(N^T\mathbf{x} + G^T\boldsymbol{\lambda}) \tag{8.22}$$

How do we know the matrix B is invertible (nonsingular)? Because it is a weighting matrix that we choose and we will be certain to choose it to be invertible! Next, we substitute $\mathbf{u}$ into Eqs. (8.20b) and (8.21), resulting in

$$\dot{\boldsymbol{\lambda}} = -(A - NB^{-1}N^T)\mathbf{x} - (F^T - NB^{-1}G^T)\boldsymbol{\lambda} \tag{8.23a}$$

and

$$\dot{\mathbf{x}} = (F - GB^{-1}N^T)\mathbf{x} - GB^{-1}G^T\boldsymbol{\lambda} \tag{8.23b}$$

Next, define

$$\hat{A} \equiv A - NB^{-1}N^T \tag{8.24a}$$

and

$$\hat{F} \equiv F - GB^{-1}N^T \tag{8.24b}$$

then for the $2n \times 1$ vector $\mathbf{z}$ defined by $\mathbf{z}^T \equiv [\mathbf{x}^T\boldsymbol{\lambda}^T]$, from Eqs. (8.23a, 8.23b), we have the linear system

$$\dot{\mathbf{z}} = \begin{bmatrix} \hat{F} & -GB^{-1}G^T \\ -\hat{A} & -\hat{F}^T \end{bmatrix} \mathbf{z} \tag{8.25}$$

Next, we will solve the linear system Eq. (8.25) using the *sweep method*. Because all the equations are linear, we make the reasonable assumption:

$$\boldsymbol{\lambda}(t) = S(t)\mathbf{x}(t) \tag{8.26}$$

with $S(t_f) = S_F$ from Eq. (8.20c). Once we have determined $S(t)$ we can sweep the initial state to the final time and have conditions on both the state and the adjoint vector at the final time, thus solving the 2PBVP.

From Eq. (8.26)

$$\dot{\boldsymbol{\lambda}} = \dot{S}\mathbf{x} + S\dot{\mathbf{x}} \tag{8.27}$$

and from Eqs. (8.23a) and (8.23b)

$$-\hat{A}\mathbf{x} - \hat{F}^T S\mathbf{x} = \dot{S}\mathbf{x} + S\hat{F}\mathbf{x} - SGB^{-1}G^T S\mathbf{x} \tag{8.28a}$$

where we have used $\boldsymbol{\lambda} = S\mathbf{x}$. Then it follows that

$$\mathbf{0} = (\dot{S} + S\hat{F} + \hat{F}^T S - SGB^{-1}G^T S + \hat{A})\mathbf{x} \tag{8.28b}$$

To be satisfied for arbitrary $\mathbf{x}$ it is necessary that

$$\dot{S} + S\hat{F} + \hat{F}^T S - SGB^{-1}G^T S + \hat{A} = \mathbf{O}_n \tag{8.29}$$

with $S(t_f) = S_F$.

Eq. (8.29) is a *Matrix Riccati Equation* because of the quadratic term in S.

Now that we have an equation to solve for $S(t)$ we can return to Eq. (8.25) and solve it using a transition matrix as

$$\mathbf{z}(t) = \Theta(t, t_f)\mathbf{z}(t_f); \quad \mathbf{z}(t_f) = \begin{bmatrix} \mathbf{x}(t_f) \\ S_F\mathbf{x}(t_f) \end{bmatrix} \tag{8.30}$$

Partitioning the $2n \times 2n$ transition matrix into $n \times n$ partitions

$$\Theta = \begin{bmatrix} \Theta_{xx} & \Theta_{x\lambda} \\ \Theta_{\lambda x} & \Theta_{\lambda\lambda} \end{bmatrix} \tag{8.31}$$

We can now write the solution of Eq. (8.30) as

$$\mathbf{x}(t) = \Theta_{xx}(t, t_f)\mathbf{x}(t_f) + \Theta_{x\lambda}(t, t_f)S_F\mathbf{x}(t_f) \tag{8.31a}$$

and

$$\boldsymbol{\lambda}(t) = \Theta_{\lambda x}(t, t_f)\mathbf{x}(t_f) + \Theta_{\lambda\lambda}(t, t_f)S_F\mathbf{x}(t_f) \tag{8.31b}$$

Using $\boldsymbol{\lambda}(t) - S(t)\mathbf{x}(t) = \mathbf{0}$ we write

$$[\Theta_{\lambda x} + \Theta_{\lambda\lambda}S_F - S(\Theta_{xx} + \Theta_{x\lambda}S_F)]\mathbf{x}(t_f) = \mathbf{0} \tag{8.32}$$

To be satisfied for arbitrary $\mathbf{x}(t_f)$ it is necessary that the bracket factor be an $n \times n$ zero matrix, for which

$$S(\Theta_{xx} + \Theta_{x\lambda}S_F) = \Theta_{\lambda x} + \Theta_{\lambda\lambda}S_F \tag{8.33}$$

Solving for S:

$$S(t) = \Lambda(t)X^{-1}(t) \tag{8.34}$$

where

$$\Lambda(t) \equiv \Theta_{\lambda x}(t, t_f) + \Theta_{\lambda\lambda}(t, t_f)S_F \tag{8.35a}$$

and

$$X(t) \equiv \Theta_{xx}(t, t_f) + \Theta_{x\lambda}(t, t_f)S_F \tag{8.35b}$$

8.4 Second-order NC and SC for an optimal control problem

Consider the time interval $t_o \leq t \leq t_f^*$, where t_f^* is either a specified value of the final time or a value implicitly determined by a terminal constraint. In addition to the first-order NC discussed in Chapter 3 there are two second-order NC: the matrix $H_{uu} \geq 0$ (is positive semidefinite, the *Legendre–Clebsch Condition*) and the solution $S(t)$ to the Matrix Riccati Equation (8.29) is finite. That $S(t)$ be finite is the *Jacobi Condition,* also known as the Jacobi No-Conjugate-Point Condition. Note that this is a NC in addition to the first-order conditions of Chapter 3 and a solution that satisfies all the first-order NC is nonoptimal if it violates the Jacobi Condition. Related SC are that $H_{uu} > 0$ (is positive definite, the *Strengthened Legendre–Clebsch Condition*) and $S(t)$ is finite. Note that the conditions on H_{uu} are consistent with the Minimum Principle.

Requiring $S(t)$ to be finite involves testing for a so-called *conjugate point.* If a conjugate point exists, the solution violates the Jacobi Condition and is therefore nonoptimal. Reference 8.1 summarizes and illustrates by example a relatively recent simplified procedure for this test, which relies on Eq. (8.34). If the matrix $X(t)$ becomes singular at some value of the time, the matrix $S(t)$ becomes unbounded and the Jacobi Condition is violated (see Problem 8.3). So the test for a matrix being unbounded is replaced by a test for a scalar (the determinant of X) being zero.

8.5 Conjugate point test procedure

An additional function that is needed is defined as

$$\mathbf{\Omega}\left[\mathbf{x}(t_f), \mathbf{u}(t_f), t_f, \boldsymbol{\nu}\right] \equiv \frac{d\Phi}{dt_f}\left[\mathbf{x}(t_f), \mathbf{u}(t_f), t_f, \boldsymbol{\nu}\right] + L\left[\mathbf{x}(t_f), \mathbf{u}(t_f), t_f\right] \tag{8.36a}$$

with

$$\begin{aligned}\frac{d\Phi}{dt_f}\left[\mathbf{x}(t_f), \mathbf{u}(t_f), t_f, \boldsymbol{\nu}\right] &= \Phi_{t_f}\left[\mathbf{x}(t_f), t_f, \boldsymbol{\nu}\right] \\ &+ \Phi_{\mathbf{x}(t_f)}\left[\mathbf{x}(t_f), t_f, \boldsymbol{\nu}\right] \mathbf{f}\left[\mathbf{x}(t_f), \mathbf{u}(t_f), t_f\right]\end{aligned} \tag{8.36b}$$

where Φ_{t_f} and $\Phi_{\mathbf{x}(t_f)}$ represent partial derivatives of the function Φ following Eq. (8.7). In addition, one terminal constraint from $\boldsymbol{\psi}[\mathbf{x}(t_f), t_f]$, taken to be the last (or only, if $q = 0$) component ψ_{q+1}, can be used to relate a small change in t_f to a small change in the state at the optimal final time t_f^* assuming that a nontangency condition is satisfied given by

$$\frac{d\psi_{q+1}}{dt_f}\left[\mathbf{x}(t_f), \mathbf{u}(t_f), t_f\right] \neq 0 \tag{8.37}$$

This results in only q terminal constraints to be considered from the standpoint of controllability. If necessary, the constraints can be renumbered so that the last component satisfies Eq. (8.37).

The second-order test procedure described here and in Ref. [8.1] is based on Ref. [8.2] with improved notation, and on Refs. [8.3–5]. The matrix S_F in the expression for $X(t)$ in Eq. (8.35b) is given by

$$\begin{aligned} S_F = \Phi_{\mathbf{x}(t_f)\mathbf{x}(t_f)} - \Omega^T_{\mathbf{x}(t_f)}\left(\frac{d\psi_{q+1}}{dt_f}\right)^{-1}\left(\psi_{q+1}\right)_{\mathbf{x}(t_f)} - \left(\psi_{q+1}\right)^T_{\mathbf{x}(t_f)}\left(\frac{d\psi_{q+1}}{dt_f}\right)^{-1}\Phi_{\mathbf{x}(t_f)} \\ + \left(\psi_{q+1}\right)^T_{\mathbf{x}(t_f)}\left(\frac{d\psi_{q+1}}{dt_f}\right)^{-1}\left[\Omega_{t_f} + \Omega_{\mathbf{x}(t_f)}\mathbf{f}\right]\left(\frac{d\psi_{q+1}}{dt_f}\right)^{-1}\left(\psi_{q+1}\right)_{\mathbf{x}(t_f)} \end{aligned} \tag{8.38}$$

Note that if the last (or only, if $q = 0$) terminal constraint ψ_{q+1} specifies the value of the final time, only the first term on the right-hand side of Eq. (8.38) is nonzero, because all the partial derivatives of ψ_{q+1} with respect to the final state are zero. Also note that from Eqs. (8.35b) and (8.14c) the final value $X(t_f) = \mathbf{I}_n$.

The test procedure described in Ref. [8.1] is different for the case of a single terminal constraint ($q = 0$) and multiple terminal constraints ($q > 0$). In both cases a smooth (no corners) trajectory $\mathbf{x}^*(t)$ is assumed, and this requires a continuous control $\mathbf{u}^*(t)$.

8.5.1 Single terminal constraint ($q = 0$)

Let a continuous control function $\mathbf{u}^*(t)$ for $t_o \leq t \leq t_f^*$ be a stationary solution to the optimal trajectory problem. Denote the state vector, adjoint vector, and final time on the stationary solution as $\mathbf{x}^*(t)$, $\boldsymbol{\lambda}^*(t)$, and t_f^*, respectively. NC (in addition to being a stationary solution) are that the $m \times m$ matrix $H_{\mathbf{uu}}\left[\mathbf{x}^*(t), \mathbf{u}^*(t), \boldsymbol{\lambda}^*(t), t\right]$ be positive semidefinite and that $\det[X(t)]$ in Eq. (8.35b) be nonzero for $t_o \leq t \leq t_f^*$.

SC are that if H_{uu} is positive definite for $t_o \leq t \leq t_f^*$ and if $\det[X(t)]$ is nonzero for $t_o \leq t \leq t_f^*$,then the stationary solution furnishes a weak local minimum of the cost. The conditions on H_{uu} are the classical Legendre–Clebsch conditions and the nonzero $\det[X(t)]$ is the second-order Jacobi no-conjugate-point condition.

The Jacobi condition is both a NC and part of the SC so that if $\det[X(t_c)] = 0$ a conjugate point exists at time t_c in the case of a single terminal constraint. If this occurs, the stationary solution is nonoptimal, and there exists a neighboring trajectory of lower cost.

8.5.2 Multiple terminal constraints ($q > 0$)

For multiple terminal constraints the second-order conditions can be more complicated and may require calculation of two matrices: $X(t)$ from Eq. (8.35b) and another matrix called $\hat{X}(t)$. Let a continuous control function $\mathbf{u}^*(t)$ for $t_o \leq t \leq t_f^*$ be a stationary solution to the optimal trajectory problem. NC (in addition to being a stationary solution) are that the $m \times m$ matrix $H_{uu}\left[\mathbf{x}^*(t), \mathbf{u}^*(t), \boldsymbol{\lambda}^*(t), t\right]$ be positive semidefinite for $t_o \leq t \leq t_f^*$, and that there exists a time t_1 with $t_o \leq t_1 < t_f^*$ such that $\det[X(t)]$ is nonzero for $t_1 \leq t \leq t_f^*$, and $\det[\hat{X}(t)]$ (defined below) is nonzero for $t_o \leq t \leq t_1$.

SC are that if H_{uu} is positive definite for $t_o \leq t \leq t_f^*$, and if $\det[X(t)]$ is nonzero for $t_1 \leq t \leq t_f^*$, and $\det[\hat{X}(t)]$ is nonzero for $t_o \leq t \leq t_1$, then the stationary solution furnishes a weak local minimum of the cost.

The selection of the time t_1 is described in Section 8.6 Computational Procedure. If $\det[\hat{X}(t_c)] = 0$, the Jacobi NC is violated and a conjugate point exists at time t_c in the case of multiple constraints. If this occurs, the stationary solution is nonoptimal and there exists a neighboring solution of lower cost.

The matrix $X(t)$ is calculated for $t_1 \leq t \leq t_f^*$ using Eq. (8.35b). The matrix $\hat{X}(t)$ is defined in an analogous way for $t_o \leq t \leq t_1$ by

$$\hat{X}(t) = \Theta_{\mathbf{xx}}(t, t_1) + \Theta_{\mathbf{x}\boldsymbol{\lambda}}(t, t_1)\hat{S}(t_1) \tag{8.39}$$

By analogy with Eq. (8.14c), $\Theta(t_1, t_1) = \mathbf{I}_{2n}$ and $\hat{X}(t_1) = \mathbf{I}_n$.

The boundary condition $\hat{S}(t_1)$ is calculated as (Ref. [8.1])

$$\hat{S}(t_1) = S(t_1) - R(t_1)Q^{-1}(t_1)R^T(t_1) \tag{8.40}$$

where

$$S(t_1) = \Lambda(t_1)X^{-1}(t_1) \tag{8.41}$$

and

$$\Lambda(t) = \Theta_{\boldsymbol{\lambda}\mathbf{x}}(t, t_f) + \Theta_{\boldsymbol{\lambda}\boldsymbol{\lambda}}(t, t_f)S_f \tag{8.42}$$

The $n \times q$ matrix $R(t)$ is calculated from

$$\dot{R}(t) = [\Lambda(t)X^{-1}(t)A_2(t) - A_1^T(t)]R(t) \tag{8.43a}$$

with boundary condition

$$R(t_f) = \bar{\psi}^T_{\mathbf{x}(t_f)} - [(\psi_{q+1})_{\mathbf{x}(t_f)}]^T \left(\frac{d\psi_{q+1}}{dt_f}\right)^{-1} \left(\frac{d\bar{\psi}}{dt_f}\right)^T \tag{8.43b}$$

where the vector $\bar{\psi}$ represents the first q components of the vector ψ.

The $q \times q$ matrix $Q(t)$ is calculated from

$$\dot{Q}(t) = R^T(t)A_2(t)R(t) \tag{8.44a}$$

with boundary condition

$$Q(t_f) = \mathbf{O}_q \tag{8.44b}$$

8.6 Computational procedure

A summary of the procedure and computational steps follows. This procedure is applied to a stationary trajectory that satisfies all the first-order NC.

1. Determine whether the matrix H_{uu} is positive definite for $t_o \leq t \leq t_f^*$.
2. Calculate the matrices $A_o(t)$, $A_1(t)$, and $A_2(t)$ from Eqs. (8.11b, 8.11c, and 8.12b).
3. Form the matrix $P(t)$ in Eq. (8.13), and integrate Eq. (8.14b) backwards from the final time t_f^* to t_o using boundary condition (8.14b) to calculate the matrix $\Theta(t, t_f)$ in Eq. (8.31). This is shown in Fig. 8.1.
4. Simultaneously calculate det$[X(t)]$, by Eq. (8.35b) using the matrix S_F from Eq. (8.38).

The remainder of the procedure depends on whether the terminal constraint is single ($q = 0$) or multiple ($q > 0$). The remaining condition for a single terminal constraint

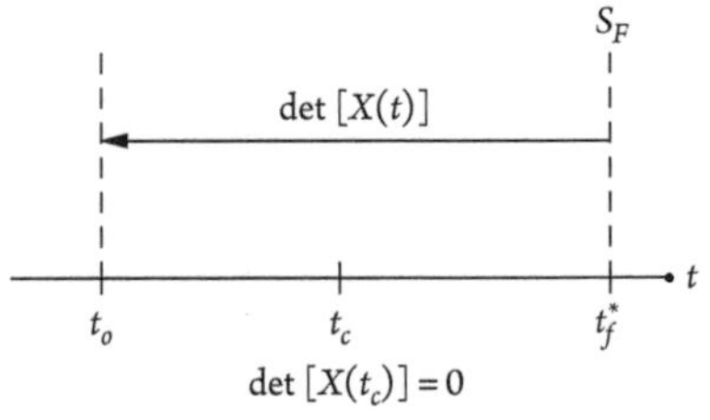

Figure 8.1 Calculation of det$[X(t)]$ for $q = 0$

is denoted by 5S, and the remaining conditions for multiple terminal constraints are denoted by 5M and 6M.

5S. For $q = 0$, if there exists a time t_c with $t_o \leq t_c \leq t_f^*$ for which $\det[X(t_c)] = 0$, then a conjugate point exists at time t_c and the trajectory is nonoptimal. This is shown in Fig. 8.1. But if $\det[X(t)]$ is nonzero for $t_o \leq t \leq t_f^*$ then the stationary solution furnishes a weak local minimum of the cost. The procedure for a single terminal constraint is completed at this point.

5M. For $q > 0$, initially select the time t_1 to be the initial time t_o. If $\det[X(t)]$ is nonzero for $t_o \leq t \leq t_f^*$, then the stationary solution furnishes a weak local minimum of the cost. The procedure for multiple terminal constraints is completed at this point. But if there exists a time t_z with $t_o \leq t_z \leq t_f^*$ for which $\det[X(t_z)] = 0$, then a conjugate point *may* exist. See Fig. 8.2a.

Further testing is required as described in Step 6M.

6M. Select a new time t_1 for which $t_z < t_1 < t_f^*$. Then $\det[X(t)]$ will be nonzero for $t_1 \leq t \leq t_f^*$. Calculate $\det[\hat{X}(t)]$ for $t_o \leq t \leq t_1$ using Eq. (8.39) and other matrices defined by Eqs. (8.40–8.44b). If $\det[\hat{X}(t_c)] = 0$ for $t_o \leq t_c \leq t_1$, a conjugate point exists at time t_c and the trajectory is nonoptimal. This is shown in Fig. 8.2b. But if $\det[\hat{X}(t)]$ is nonzero for $t_o \leq t \leq t_1$, then the stationary solution furnishes a weak local minimum of the cost. The procedure for multiple terminal constraints is completed at this point.

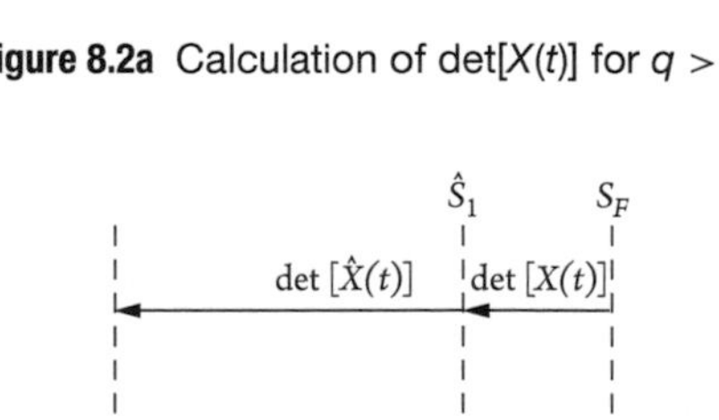

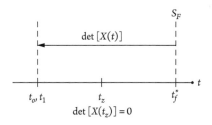

Figure 8.2a Calculation of $\det[X(t)]$ for $q > 0$

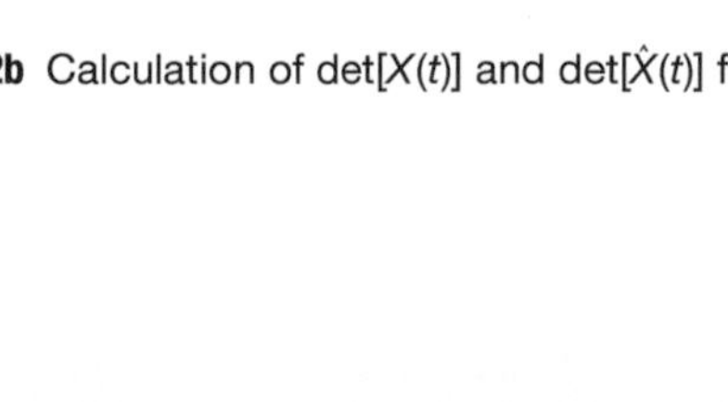

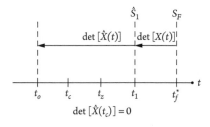

Figure 8.2b Calculation of $\det[X(t)]$ and $\det[\hat{X}(t)]$ for $q > 0$

Example 8.2 Rigid body control

Let's apply the second-order conditions to Example 3.3 Rigid Body Control. In that example $n = m = 1$ and there is no terminal constraint. But in the formalism of this chapter the specified value of the final time T is a terminal constraint and $q = 0$, so

$$\begin{aligned}
&\psi = t_f^* - T = 0 \\
&H = \frac{1}{2}u^2 + \lambda u, \text{ so} \\
&H_{uu} = 1 > 0 \quad \text{and} \quad H_{xx} = H_{ux} = H_{xu} = 0 \\
&\dot{x} = f = u, \text{ so } f_x = 0 \quad \text{and} \quad f_u = 1 \\
&A_o = H_{xx} - H_{xu}H_{uu}^{-1}H_{ux} = 0, \\
&A_1 = f_x - f_u H_{uu}^{-1}H_{ux} = 0, \\
&A_2 = f_u H_{uu}^{-1} f_u^T = 1.
\end{aligned}$$

The P matrix is then

$$P = \begin{bmatrix} 0 & -1 \\ 0 & 0 \end{bmatrix} \tag{8.45}$$

and the corresponding transition matrix is

$$\Theta(t - T) = \begin{bmatrix} 1 & T - t \\ 0 & 1 \end{bmatrix} \tag{8.46}$$

Because $\Phi = \frac{1}{2}kx^2(T) + \nu(t_f^* - T)$, $S_F = \Phi_{x(T)x(T)} = k$, for which $X(t) = 1 + k(T - t)$ which yields $t_c = \frac{1+kT}{k} = \frac{1}{k} + T$.

Assuming $k > 0$, $t_c > T$ which is outside the defined time interval, so no conjugate point exists and therefore the stationary solution is a local minimum.

Example 8.3

Example 3 in Ref. [8.2] is an optimal variable-specific-impulse trajectory. What appears here is merely a summary; all the details are in the reference. It is a planar problem and the equations of motion are described in polar coordinates. In terms of the $n = 3$ state variables $\mathbf{x}^T = [r \; \mathrm{v}_r \; \mathrm{v}_\theta]$ The equations of motion are

$$\dot{r} = \mathrm{v}_r \tag{8.47a}$$

$$\dot{\mathrm{v}}_r = \mathrm{v}_\theta^2/r - \mu/r^2 + \Gamma \sin\beta \tag{8.47b}$$

$$\dot{\mathrm{v}}_\theta = -\mathrm{v}_r \mathrm{v}_\theta/r + \Gamma\cos\beta \tag{8.47c}$$

where Γ is the magnitude of the thrust acceleration and β is the thrust angle relative to the local horizontal. (The polar angle can be calculated using $\dot{\theta} = v_\theta / r$ but θ is not needed as a state variable.)

The only terminal constraint is a specified value of the final time, so $q = 0$. The objective of the problem is to increase the total energy of the orbit with a penalty on fuel consumption. A composite cost functional to be minimized is defined that is a linear combination of the negative of the total energy at the final time t_f and the fuel consumed:

$$J = -E(t_f) + \kappa \int_{t_o}^{t_f} \frac{\Gamma^2(t)}{2} dt \tag{8.48}$$

In Eq. (8.48) κ is a positive weighting factor for the fuel consumption term given by Eq. (4.23).

To calculate the amount of fuel consumed a new variable α is defined by

$$\dot{\alpha} = \mathbf{\Gamma}^2/2 \tag{8.49}$$

and Eq. (8.49) is integrated along with the state equations (8.47a–c).

The first-order NC are used to determine the two control variables $\beta(t)$ and $\Gamma(t)$ on the stationary solution. A canonical distance unit (DU) and time unit (TU) are used, for which $\mu = 1\mathrm{DU}^3/\mathrm{TU}^2$.

To test the second-order conditions on the stationary solution a final state is arbitrarily chosen to be a circular orbit of radius 10 DU. The final state vector is then $\mathbf{x}^T(t_f) = [r \ v_r \ v_\theta] = [10 \ 0 \ 1/\sqrt{10}]$ for which the value of the final energy is $E(t_f) = -0.05$. The final time t_f^* is taken to be 95 TU and the weighting factor in the cost is $\kappa = 100$, which gives approximately equal weight to the energy and fuel terms in the cost.

All the relevant equations including the first-order NC are integrated backwards from the final time of 95 TU to $t = 0$. The resulting initial state is

$$\mathbf{x}^T(0) = \begin{bmatrix} 0.8209 & 0.0210 & 1.4142 \end{bmatrix} \tag{8.50}$$

and for $\alpha(95) = 0$, $\alpha(0) = -0.000732$. This is equivalent to $\alpha(95) = 0.000732$ for $\alpha(0) = 0$. The cost in Eq. (8.48) is then

$$J = -E(95) + \kappa\alpha(95) = 0.05 + 100 \times 0.000732 = 0.1232 \tag{8.51}$$

Next, the second-order conditions are applied to this stationary trajectory. In terms of the control variables $\mathbf{u}^T = [\,\Gamma \ \beta\,]$:

$$H_{uu} = \begin{bmatrix} \kappa & 0 \\ 0 & \kappa\Gamma^2 \end{bmatrix} \tag{8.52}$$

Table 8.1 Trajectory comparisons

Variable	Stationary Trajectory	GA Trajectory
Negative of final energy	0.0500	0.0294
Weighted amount of fuel	0.0732	0.0763
Cost	0.1232	0.1057

which is positive definite and the Strengthened Legendre–Clebsch condition is satisfied. Calculation of the 3×3 matrix $X(t)$ results in $\det[X(t)] = 0$ at $t = 38.6$ TU. Because there is a single terminal constraint this means that the stationary solution contains a conjugate point at $t_c = 38.6$ TU and is nonoptimal. There exists a neighboring solution of lower cost.

A lower cost solution is found using a genetic algorithm (GA) starting from the initial state given by Eq. (8.50) and running time forward to 95 TU. Table 8.1 shows the results for the GA trajectory and compares them to the stationary trajectory.

Note that in Table 1 the cost of the GA trajectory is 14% less than the nonoptimal stationary trajectory. The GA solution has a higher final orbit energy (–0.0294) than the stationary solution (–0.0500) but uses a little more fuel to achieve the higher final energy. The qualitative differences in the solutions and an interpretation is discussed in Ref. [8.2].

Another example of a conjugate point is presented in more detail Appendix G.

Problems

8.1 a) Determine if $f(x) = x^3$ satisfies the SC of Eq. (8.3).

b) Based on Eq. (8.4) determine what type of stationary point $x^* = 0$ is (maximum, minimum, etc.).

8.2 Verify that for $S(t)$ in Eq. (8.34) $S(t) \to S_F$ as $t \to t_f$.

8.3 Explain why a singular matrix X results in an unbounded matrix S. *Hint*: the determinant of the product of two square matrices is equal to the product of their two determinants.

8.4 Discuss the existence of a conjugate point in Example 8.2 for the case $k < 0$.

References

[8.1] Prussing, J.E. and Sandrik, S.L., "Second-Order Necessary Conditions and Sufficient Conditions Applied to Continuous-Thrust Trajectories", *Journal of Guidance, Control, and Dynamics*, Vol. 28, No. 4, Jul–Aug 2005, pp. 812–16.

[8.2] Jo, J-W., and Prussing, J.E., "Procedure for Applying Second-Order Conditions in Optimal Control Problems", *Journal of Guidance, Control, and Dynamics*, Vol. 23, No. 2, Mar–Apr 2000, pp. 241–50.

[8.3] Wood, L.J., "Sufficient Conditions for a Local Minimum of the Bolza Problem with Multiple Terminal Point Constraints", American Astronautical Society, Paper 91–450, Aug 1991.

[8.4] Wood, L.J., "Sufficient Conditions for a Local Minimum of the Bolza Problem with a Scalar Terminal Point Constraint", *Advances in the Astronautical Sciences*, edited by B. Kaufman, K. Alfriend, R. Roehrich, and R. Dasenbrock, Vol. 76, Part 3, Univelt, Inc., San Diego, CA, pp. 2053–72; also American Astronautical Society, Paper 91–449, Aug. 1991.

[8.5] Breakwell, J.V., and Ho, Y-C., "On the Conjugate Point Condition for the Control Problem", *International Journal of Engineering Science*, Vol 2, 1965, pp. 565–79.

[8.6] Prussing, J.E., "Simplified Conjugate Point Procedure", *Journal of Guidance, Control, and Dynamics*, Vol. 40, No. 5, 2017, pp. 1255–57.

Appendix A
Lagrange Multiplier Interpretation

Consider the problem of minimizing $L(\mathbf{x}, \mathbf{u})$ subject to the constraint equations:

$$\mathbf{f}(\mathbf{x}, \mathbf{u}, \mathbf{c}) = \mathbf{0} \tag{A.1}$$

where $\mathbf{f}$ and $\mathbf{x}$ are n-dimensional, $\mathbf{u}$ is m-dimensional, and $\mathbf{c}$ is a q-dimensional vector of constants in the constraint equations.

The differential change in the vector $\mathbf{f}$ is given by:

$$d\mathbf{f} = \mathbf{f}_x d\mathbf{x} + \mathbf{f}_u d\mathbf{u} + \mathbf{f}_c d\mathbf{c} \tag{A.2}$$

where $\mathbf{f}_c$ is an $n \times q$ Jacobian matrix.

Considering differential changes in variables from their stationary values, we wish to hold the constraint $\mathbf{f} = \mathbf{0}$ by requiring $d\mathbf{f} = \mathbf{0}$. This results in

$$d\mathbf{x} = -\mathbf{f}_x^{-1}(\mathbf{f}_u d\mathbf{u} + \mathbf{f}_c d\mathbf{c}) \tag{A.3}$$

Starting with

$$dL = L_x d\mathbf{x} + L_u d\mathbf{u} \tag{A.4}$$

which, using Eq. (A.3), is equal to

$$dL = (L_u - L_x \mathbf{f}_x^{-1} \mathbf{f}_u) d\mathbf{u} - L_x \mathbf{f}_x^{-1} \mathbf{f}_c d\mathbf{c} \tag{A.5}$$

The NC that $H_x = \mathbf{0}^T$ results in $\boldsymbol{\lambda}^T = -L_x \mathbf{f}_x^{-1}$ which, substituted into $H_u = \mathbf{0}^T$ yields:

$$L_u - L_x \mathbf{f}_x^{-1} \mathbf{f}_u = \mathbf{0}^T \tag{A.6}$$

Substituting into Eq. (A.5) yields:

$$dL = \boldsymbol{\lambda}^T \mathbf{f}_c d\mathbf{c} \tag{A.7}$$

from which

$$\frac{\partial L}{\partial \mathbf{c}} = \boldsymbol{\lambda}^T \mathbf{f}_c \tag{A.8}$$

The gradient $\frac{\partial L}{\partial \mathbf{c}}$ is an $1 \times q$ row vector. Each element represents the *sensitivity* of the stationary value of the cost L to a small change in the corresponding component of the vector $\mathbf{c}$, i.e., the change in a particular constant in the constraint equation.

In component form,

$$dL = \lambda_j \frac{\partial f_j}{\partial c_k} dc_k \tag{A.9}$$

or

$$\frac{\partial L}{\partial c_k} = \lambda_j \frac{\partial f_j}{\partial c_k} \tag{A.10}$$

Note that all values above are evaluated at the stationary point.

Example A.1

As a hypothetical example consider the terminal constraint as the ellipse

$$f(x, y) = \frac{x^2}{a^2} + \frac{y^2}{b^2} - 1 = 0 \tag{A.11}$$

with nominal values $a = 2$ and $b = 1$. Suppose a numerical solution yields $x = 1.5$, $L = 1.1$, and $\lambda = 1.7$.

Consider a small increase of 0.1 in the value of a. Because $f_a = -2x^2/a^3$ the corresponding change in the value of L is

$$dL = \lambda f_a da = (1.7)(-4.5/8)(0.1) = -0.0956 \tag{A.12}$$

so to first order the value of L decreases from 1.1 to 1.0044. Note that we obtained this result without knowing the expression for the cost function.

Appendix B
Hohmann Transfer

In this appendix two very different derivations are presented for the optimality of the Hohmann transfer. In Section B.1 the problem is treated as a constrained parameter optimization, treated in Section 1.2 using Lagrange multipliers. In Section B.2 ordinary gradients are used.

B.1 Constrained parameter optimization

A demonstration that the Hohmann transfer provides a stationary value of the total Δv for a coplanar two-impulse circle-to-circle transfer is given in Ref. [B.1]. The notation used here is slightly different.

The velocity changes due to the thrust impulses are described in terms of their transverse and radial components Δv_θ and Δv_r. The initial circular orbit velocity at radius r_1 is v_{c1} and the final circular orbit velocity at r_2 is v_{c2}. The two velocity changes Δv_1 and Δv_2 are written in terms of their transverse and radial components as

$$\Delta v_{1\theta} = v_1 \cos \gamma_1 - v_{c1} \quad \Delta v_{1r} = v_1 \sin \gamma_1 - 0 \tag{B.1a}$$

$$\Delta v_{2\theta} = v_{c2} - v_2 \cos \gamma_2 \quad \Delta v_{2r} = 0 - v_2 \sin \gamma_2 \tag{B.1b}$$

where γ_k are the path angles of the velocity changes (with respect to the local horizontal) and v_k are the transfer orbit velocity magnitudes at the terminal radii r_1 and r_2. In Eqs. (B.1a–b) we have assumed $r_2 > r_1$ (inner to outer orbit) so that the $\Delta v_{i\theta}$ components are positive and equal to their magnitudes. If $r_2 < r_1$ the directions of all the Δvs are reversed with the magnitudes unchanged.

The magnitudes of the velocity changes are then

$$\Delta v_1 = \sqrt{(v_1 \cos \gamma_1 - v_{c1})^2 + (v_1 \sin \gamma_1)^2} = \sqrt{(v_1 - v_{c1})^2 + 2v_{c1}v_1(1 - \cos \gamma_1)} \tag{B.2a}$$

$$\Delta v_2 = \sqrt{(v_{c2} - v_2 \cos \gamma_2)^2 + (v_2 \sin \gamma_2)^2} = \sqrt{(v_{c2} - v_2)^2 + 2v_2v_{c2}(1 - \cos \gamma_2)} \tag{B.2b}$$

The total velocity change is then

$$\Delta v_{tot} = \Delta v_1 + \Delta v_2 \tag{B.3}$$

There are two equality constraints given by the expressions for conservation of energy and angular momentum on the transfer orbit. For conservation of energy

$$v_1^2/2 - \mu/r_1 = v_2^2/2 - \mu/r_2 \tag{B.4}$$

where μ is the gravitational constant (Ref. [B.4]). Using the equations for circular orbit speed $v_{c1}^2 = \mu/r_1$ and $v_{c2}^2 = \mu/r_2$ to eliminate r_1 and r_2 in Eq. (B.4) results in the energy constraint equation

$$v_1^2 - v_2^2 - 2(v_{c1}^2 - v_{c2}^2) = 0 \tag{B.5}$$

For conservation of angular momentum

$$r_1 v_1 \cos\gamma_1 = r_2 v_2 \cos\gamma_2 \tag{B.6}$$

Eliminating r_1 and r_2 as above results in the angular momentum constraint equation

$$v_1 v_{c2}^2 \cos\gamma_1 - v_2 v_{c1}^2 \cos\gamma_2 = 0 \tag{B.7}$$

The constrained minimization problem is then to minimize the cost in Eq. (B.3) with the Δv_k given by Eqs. (B.2) subject to the constraints of Eqs. (B.5) and (B.7).

Letting $c_k \equiv \cos\gamma_k$ and $s_k \equiv \sin\gamma_k$ use Eq. (1.10) from Chapter 1 to form the function

$$H = \Delta v_1 + \Delta v_2 + \lambda_e[v_1^2 - v_2^2 + 2(v_{c1}^2 - v_{c2}^2)] + \lambda_m(v_1 v_{c2}^2 c_1 - v_2 v_{c1}^2 c_2) \tag{B.8}$$

There are four variables: v_1 v_2, γ_1, γ_2, along with the multipliers λ_e and λ_m. The first-order NC (1.11) are

$$\frac{\partial H}{\partial \gamma_1} = \frac{\partial \Delta v_1}{\partial \gamma_1} + \frac{\partial \Delta v_2}{\partial \gamma_1} - \lambda_m v_1 v_{c2}^2 s_1 = 0 \tag{B.9a}$$

$$\frac{\partial H}{\partial \gamma_2} = \frac{\partial \Delta v_1}{\partial \gamma_2} + \frac{\partial \Delta v_2}{\partial \gamma_2} + \lambda_m v_2 v_{c1}^2 s_2 = 0 \tag{B.9b}$$

$$\frac{\partial H}{\partial v_1} = \frac{\partial \Delta v_1}{\partial v_1} + \frac{\partial \Delta v_2}{\partial v_1} + 2\lambda_e v_1 + \lambda_m v_{c2}^2 c_1 = 0 \tag{B.9c}$$

$$\frac{\partial H}{\partial v_2} = \frac{\partial \Delta v_1}{\partial v_2} + \frac{\partial \Delta v_2}{\partial v_2} - 2\lambda_e v_2 - \lambda_m v_{c1}^2 c_2 = 0 \tag{B.9d}$$

and the constraint equations (B.5) and (B.7),

The reader can verify that

$$\frac{\partial \Delta v_1}{\partial \gamma_1} = \frac{v_1 v_{c1} s_1}{\Delta v_1}; \qquad \frac{\partial \Delta v_2}{\partial \gamma_1} = 0 \tag{B.10a}$$

$$\frac{\partial \Delta v_1}{\partial \gamma_2} = 0; \qquad \frac{\partial \Delta v_2}{\partial \gamma_2} = \frac{v_2 v_{c2} s_2}{\Delta v_2} \tag{B.10b}$$

Then Eq. (B.9a) becomes

$$\left[\frac{v_1 v_{c1}}{\Delta v_1} - \lambda_m v_1 v_{c2}^2\right] s_1 = 0 \tag{B.11a}$$

and Eq. (B.9b) becomes

$$\left[\frac{v_2 v_{c2}}{\Delta v_2} + \lambda_m v_2 v_{c1}^2\right] s_2 = 0 \tag{B.11b}$$

Eqs. (B.11a) and (B.11b) are satisfied by $s_1 = s_2 = \gamma_1 = \gamma_2 = 0$, i.e., Δv_1 and Δv_2 are tangent to the terminal circular orbits, representing the Hohmann transfer.

With $c_1 = c_2 = 1$ the reader can verify that

$$\frac{\partial \Delta v_1}{\partial v_1} = 1; \qquad \frac{\partial \Delta v_2}{\partial v_1} = 0 \tag{B.12a}$$

$$\frac{\partial \Delta v_1}{\partial v_2} = 0; \qquad \frac{\partial \Delta v_2}{\partial v_2} = 1 \tag{B.12b}$$

Then Eqs. (B.9c–d) simplify to

$$1 + 2\lambda_e v_1 + \lambda_m v_{c2}^2 = 0 \tag{B.13a}$$

$$1 - 2\lambda_e v_2 - \lambda_m v_{c1}^2 = 0 \tag{B.13b}$$

which can be solved to yield

$$\lambda_e = \frac{v_{c2}^2 - v_{c1}^2}{2(v_1 v_{c1}^2 - \nu_2 \nu_{c2}^2)} \tag{B.14a}$$

$$\lambda_m = \frac{v_2 - v_1}{(v_1 v_{c1}^2 - \nu_2 \nu_{c2}^2)} \tag{B.14b}$$

Because we have assumed $r_2 > r_1$, it follows that $v_{c1} > v_{c2}$ and $v_1 > v_2$. From Eqs. (B.14a–b) we see that $\lambda_e < 0$ and $\lambda_m < 0$.

From the constraint equations (B.5) and (B.7) we can show that

$$v_1 = v_{c1}\left(\frac{2v_{c1}^2}{v_{c1}^2 + v_{c2}^2}\right)^{\frac{1}{2}} \quad \text{and} \quad v_2 = \frac{v_{c2}^2}{v_{c1}}\left(\frac{2v_{c1}^2}{v_{c1}^2 + v_{c2}^2}\right)^{\frac{1}{2}} \tag{B.15}$$

And lastly, we solve for the velocity changes in Eqs. (B.2a–b), eliminating the terminal circular orbit speeds in favor of the terminal radii using the expressions following Eq. (B.4). In terms of $R \equiv r_2/r_1 > 1$ we have the familiar nondimensional formulas

$$\frac{\Delta v_1}{v_{c1}} = \left(\frac{2R}{1+R}\right)^{\frac{1}{2}} - 1 \tag{B.16a}$$

$$\frac{\Delta v_2}{v_{c1}} = \frac{1}{\sqrt{R}}\left(1 - \left(\frac{2}{1+R}\right)^{\frac{1}{2}}\right) \tag{B.16b}$$

B.2 Simple proof of global optimality

In this section optimality is demonstrated using the analysis in Ref. [B.4], which is based on an earlier analysis in Ref. [B.2]. Reference [B.2] was inspired by Ref. [B.3].

A two-impulse transfer ellipse, described by its parameter p and eccentricity e (Ref. [B.4]), must intersect both terminal circular orbits, of radii r_1 and r_2. We will assume $r_2 > r_1$. This means that the periapse radius r_p of the transfer orbit must lie at or inside the initial radius r_1:

$$r_p = \frac{p}{(1+e)} \leq r_1 \tag{B.17a}$$

and the apoapse radius r_a must lie at or outside the final radius r_2:

$$r_a = \frac{p}{(1-e)} \geq r_2 \tag{B.17b}$$

From these two conditions we get

$$p \leq r_1(1+e) \tag{B.18a}$$

and

$$p \geq r_2(1-e) \tag{B.18b}$$

Condition (B.18a) and (B.18b) describe the region of the (p, e) plane in which the transfer orbits exist, as shown in Fig. B.1.

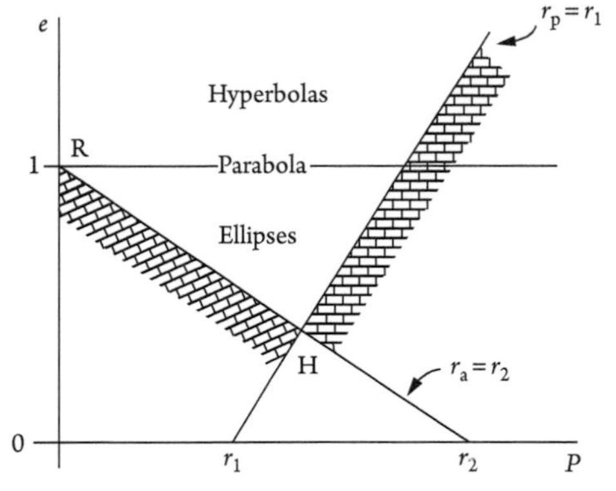

Figure B.1 Feasible transfer orbits

The region of feasible transfer orbits in Fig. B.1 contains ellipses for $0 < e < 1$, parabolas for $e = 1$, and hyperbolas for $e > 1$, with *all* the rectilinear orbits located at point R for which $e = 1$ and $p = 0$.

Point H in Fig. B.1 is interesting, if only because it represents the least eccentric (most circular) of all the feasible transfer orbits. But as we will see, point H is also the Hohmann transfer, the absolute minimum-fuel two-impulse transfer between circular orbits.

The magnitude of the velocity change to depart or enter a circular orbit at $r = r_k$, where $k = 1, 2$ is described by the law of cosines as

$$(\Delta \mathrm{v})^2 = \mathrm{v}^2 + \mathrm{v}_c^2 - 2\mathrm{v}_c \mathrm{v}_\theta \tag{B.19}$$

where v_θ is the component of the velocity vector normal to the radius and is given by $\frac{(\mu p)^{\frac{1}{2}}}{r}$. In addition $\mathrm{v}_c^2 = \frac{\mu}{r}$ and, using $p = a(1 - e^2)$, where a is the semimajor axis, the vis-viva equation (Ref. [B.4]) is

$$\mathrm{v}^2 = \mu\left(\frac{2}{r} - \frac{1}{a}\right) = \mu\left(\frac{2}{r} + \frac{e^2 - 1}{p}\right) \tag{B.20}$$

The total velocity change is then

$$\Delta \mathrm{v}_T = \Delta \mathrm{v}_1 + \Delta \mathrm{v}_2$$

where $\Delta \mathrm{v}_1$ occurs at $r = r_1$ and $\Delta \mathrm{v}_2$ occurs at $r = r_2$. Thus $\Delta \mathrm{v}$ in Eq. (B.19) can be replaced by $\Delta \mathrm{v}_k$ corresponding to $r = r_k$. If one examines the gradient of the total velocity change with respect to the eccentricity of the transfer orbit, one obtains

$$\frac{\partial \Delta \mathrm{v}_T}{\partial e} = \frac{\partial \Delta \mathrm{v}_1}{\partial e} + \frac{\partial \Delta \mathrm{v}_2}{\partial e} \tag{B.21}$$

and, from differentiating Eq. (B.19) using (B.20):

$$2\Delta v_k \frac{\partial \Delta v_k}{\partial e} = 2v_k \frac{\partial v_k}{\partial e} = \frac{2e\mu}{p.} \tag{B.22}$$

Thus

$$\frac{\partial \Delta v_T}{\partial e} = \frac{e\mu}{p}(\Delta v_1^{-1} + \Delta v_2^{-1}) > 0 \tag{B.23}$$

The fact that the gradient of the total velocity change with respect to the eccentricity is *positive* means that in Fig. B.1 at *any* point (p, e) in the interior of the feasible region the total velocity change can be *decreased* by *decreasing* the value of e while holding the value of p constant. This means that the minimum total velocity change solution will always lie on the cross-hatched boundary of the feasible region given by $r_a = r_2$ or $r_p = r_1$ or both.

What will be demonstrated is that, when the velocity change is restricted to the boundary, the gradient with respect to e remains positive. Thus the optimal solution lies at the minimum value of e on the boundary, i.e. at point H in Fig. B.1, which represents the Hohmann transfer.

To do this, the expression for the total velocity change is restricted to the boundary by substituting for the value of p along each portion of the boundary, namely $p = r_2(1-e)$ on the left part $(r_a = r_2)$ and $p = r_1(1+e)$ on the right part $(r_p = r_1)$. Thus, the total velocity change becomes a function of a *single* variable e. Letting $\Delta\hat{v}$ denote the velocity change restricted to the boundary, Eq. (B.19) becomes for the left part:

$$(\Delta\hat{v})^2 = \mu\left(\frac{2}{r} + \frac{e^2-1}{r_2(1-e)}\right) + \frac{\mu}{r} - 2\left(\frac{\mu}{r}\right)^{\frac{1}{2}}\left(\frac{\mu r_2(1-e)}{r}\right)^{\frac{1}{2}} \tag{B.24}$$

Differentiating with respect to the single variable e and recalling that $R > 1$ and $0 < e < 1$:

$$2\Delta\hat{v}_1 \frac{d\Delta\hat{v}_1}{de} = \frac{\mu}{r_2}\left[R^{3/2}\frac{1}{(1-e)^{\frac{1}{2}}} - 1\right] > 0 \tag{B.25}$$

and

$$2\Delta\hat{v}_2 \frac{d\Delta\hat{v}_2}{de} = \frac{\mu}{r_2}\left[\frac{1}{(1-e)^{\frac{1}{2}}} - 1\right] > 0 \tag{B.26}$$

and thus, similar to Eq. (B.23) $\frac{d\Delta\hat{v}_T}{de} > 0$ on the left portion of the boundary.

Similarly, on the right portion of the boundary:

$$(\Delta\hat{v})^2 = \mu\left(\frac{2}{r} + \frac{e^2-1}{r_1(1+e)}\right) + \frac{\mu}{r} - 2\left(\frac{\mu}{r}\right)^{\frac{1}{2}}\left(\frac{\mu r_1(1+e)}{r}\right)^{\frac{1}{2}} \tag{B.27}$$

Differentiating, one obtains:

$$2\Delta\hat{v}_1\frac{d\Delta\hat{v}_1}{de} = \frac{\mu}{r_1}\left[1 - \frac{1}{(1+e)^{\frac{1}{2}}}\right] > 0 \tag{B.28}$$

and

$$2\Delta\hat{v}_2\frac{d\Delta\hat{v}_2}{de} = \frac{\mu}{r_1}\left[1 - \frac{1}{R^{3/2}(1+e)^{\frac{1}{2}}}\right] > 0 \tag{B.29}$$

and thus, $\frac{d\Delta\hat{v}_T}{de} > 0$ on the right portion of the boundary also. It follows that the optimal solution lies at the point of minimum eccentricity on the boundary, namely at point H in Fig. B.1 representing the Hohmann transfer.

What has been demonstrated is that the Hohmann transfer is the absolute minimum fuel two-impulse transfer between circular orbits.

Problems

B.1 Verify Eqs. (B.10a) and (B.10b).

B.2 Verify Eqs. (B.14a) and (B.14b).

B.3 Using the interpretation of the Lagrange multiplier in Appendix A and the algebraic sign of λ_e show using Eq. (B.5) and the fact that $v_{c2}^2 = \mu/r_2$ the qualitative result that

$$\frac{\partial \Delta v_T}{\partial r_2} > 0$$

Note that this result is not immediately apparent from Eqs. (B.16a–b), but makes sense physically.

B.4 Is the result in Problem B.3 consistent with an analogous result using Eq. (B.7)?

B.5 Analogous to Prob. B.3 calculate the algebraic sign of $\frac{\partial \Delta v_T}{\partial r_1}$ and comment.

References

[B.1] Miele, A., Ciarcia, M., and Mathwig, J., "Reflections on the Hohmann Transfer", *Journal of Optimization Theory and Applications*, Vol. 123, No. 2, November 2004, pp. 233–53.

[B.2] Prussing, J.E., "Simple Proof of the Global Optimality of the Hohmann Transfer", *Journal of Guidance, Control, and Dynamics*, Vol. 15, No. 4, July–August 1992, pp. 1037–38.

[B.3] Palmore, Julian I., "An Elementary Proof of the Optimality of the Hohmann Transfer", *Journal of Guidance, Control, and Dynamics*, Vol. 7, No. 5, 1984, pp. 629–30.

[B.4] Prussing, J.E. and Conway, B.A., *Orbital Mechanics* 2nd Edition, Oxford University Press 2012.

Appendix C
Optimal Impulsive Linear Systems

In this Appendix two unrelated results for linear impulsive systems are derived. The first is that the necessary conditions (NC) of Lawden, described in Section 4.2, are also sufficient conditions. The second result is that there is a maximum number of impulses required for a solution (optimal or not) that satisfies specified boundary conditions. These results are documented in Ref. [C.1] Unfortunately, for nonlinear systems there are no corresponding results.

C.1 Sufficient conditions for an optimal solution

Consider the linear state equation

$$\dot{\mathbf{x}} = \mathbf{F}(t)\mathbf{x} \tag{C.1}$$

and the corresponding adjoint equation

$$\dot{\boldsymbol{\lambda}} = -\mathbf{F}^T(t)\boldsymbol{\lambda} \tag{C.2}$$

Using these equations

$$\frac{d}{dt}(\boldsymbol{\lambda}^T\mathbf{x}) = \dot{\boldsymbol{\lambda}}^T\mathbf{x} + \boldsymbol{\lambda}^T\dot{\mathbf{x}} = -\boldsymbol{\lambda}^T\mathbf{F}(t)\mathbf{x} + \boldsymbol{\lambda}^T\mathbf{F}(t)\mathbf{x} = 0 \tag{C.3}$$

As discussed in Appendix D the state transition matrix $\boldsymbol{\Phi}(t, \tau)$ is defined by

$$\mathbf{x}(t) = \boldsymbol{\Phi}(t, \tau)\mathbf{x}(\tau) \tag{C.4}$$

Define an adjoint transition matrix $\boldsymbol{\Psi}(t, \tau)$ analogously as

$$\boldsymbol{\lambda}(t) = \boldsymbol{\Psi}(t, \tau)\boldsymbol{\lambda}(\tau) \tag{C.5}$$

Using the result of Eq. (C.3) that $\boldsymbol{\lambda}^T(t)\mathbf{x}(t)$ is constant

$$\boldsymbol{\lambda}^T(t)\mathbf{x}(t) = \boldsymbol{\lambda}^T(\tau)\boldsymbol{\Psi}^T(t,\tau)\boldsymbol{\Phi}(t,\tau)\mathbf{x}(\tau) = \lambda^T(\tau)\mathbf{x}(\tau) \tag{C.6}$$

From Eq. (C.6) it is evident that

$$\boldsymbol{\Psi}^T(t,\tau)\boldsymbol{\Phi}(t,\tau) = \mathbf{I} \tag{C.7}$$

where **I** is the identity matrix.

From Eq. (C.7) we see that the relationship between the two transition matrices is

$$\boldsymbol{\Psi}^T(t,\tau) = \boldsymbol{\Phi}^{-1}(t,\tau) \tag{C.8a}$$

which yields from Eq. (C.5) a result we will use

$$\boldsymbol{\lambda}^T(\tau) = \boldsymbol{\lambda}^T(t)\boldsymbol{\Psi}^{-T}(t,\tau) = \boldsymbol{\lambda}^T(t)\boldsymbol{\Phi}(t,\tau) \tag{C.8b}$$

where the $-T$ superscript indicates the inverse of the transpose (equal to the transpose of the inverse).

Now consider an n-impulse transfer with each impulse providing a velocity change $\Delta\mathbf{v}_k$ at time t_k. The final state $\mathbf{x}(t_f)$ and the initial state $\mathbf{x}(t_o)$ are related by

$$\mathbf{x}(t_f) = \boldsymbol{\Phi}(t_f,t_o)\mathbf{x}(t_o) + \sum_{k=1}^{n}\boldsymbol{\Phi}(t_f,t_k)\begin{bmatrix}\mathbf{0}\\ \Delta\mathbf{v}_k\end{bmatrix} \tag{C.9}$$

Next, define the scalar c by

$$\begin{aligned} c &\equiv \lambda^T(t_f)[\mathbf{x}(t_f) - \boldsymbol{\Phi}(t_f,t_o)\mathbf{x}(t_o)] \\ &= \boldsymbol{\lambda}^T(t_f)\sum_{k=1}^{n}\boldsymbol{\Phi}(t_f,t_k)\begin{bmatrix}\mathbf{0}\\ \Delta\mathbf{v}_k\end{bmatrix} \\ &= \sum_{k=1}^{n}\boldsymbol{\lambda}^T(t_k)\begin{bmatrix}\mathbf{0}\\ \Delta\mathbf{v}_k\end{bmatrix} \end{aligned} \tag{C.10}$$

where we have used the result from Eq. (C.8b).

Considering the last three elements of $\boldsymbol{\lambda}$ to be the primer vector $\mathbf{p}$ we can write Eq. (C.10) as

$$c = \sum_{k=1}^{n}\mathbf{p}^T(t_k)\Delta\mathbf{v}_k \tag{C.11}$$

If the NC that the magnitude $p(t_k) \leq 1$ is satisfied, then

$$c = \sum_{k=1}^{n} \mathbf{p}^T(t_k)\Delta\mathbf{v}_k \leq \sum_{k=1}^{n} \Delta v_k = J \tag{C.12}$$

where J is the fuel cost, equal to the sum of the magnitudes of the velocity changes, as discussed in Section 2.3.

The equality case in Eq. (C.12) applies only if all the $\Delta\mathbf{v}_k$ are aligned with the $\mathbf{p}(t_k)$ at those instants for which $p(t_k) = 1$. These are precisely the remaining NC on the primer vector. Thus, if all the NC on the primer vector are satisfied, $J = c$; otherwise $J > c$ from Eq. (C.12), which proves that the NC are also sufficient conditions (SC) for an optimal solution.

C.2 Maximum number of impulses

In this section it is shown that a time-fixed solution (optimal or not) for a linear system requires at most q impulses, where q is equal to the number of specified final state components. A general discussion of this problem as applied to orbit transfer is given by Edelbaum in Ref. [C.2]. As an example, a three-dimensional rendezvous may require as many as six impulses, because three final position components and three final velocity components are specified. By contrast, a time-fixed coplanar intercept requires at most two impulses because two final position components are specified. A time-open intercept requires one less impulse than a time-fixed one.

The general result was shown by Neustadt in Ref. [C.3] and by Potter in Ref. [C.4]. It is the approach of Potter that is followed here. In addition to proving that no more than q impulses are required, it also provides a procedure for reducing a solution having $m > q$ impulses to a q-impulse solution without increasing the cost. The procedure reduces a $(q + 1)$-impulse to a q-impulse solution of equal or lower cost. (If $m > q + 1$ the procedure is repeated $m - (q + 1)$ times.) The resulting q impulses occur *at the same times and in the same directions as q of the original m impulses.*

Let $\mathbf{y}(t_f)$ be a q-dimensional vector containing the specified final state components, where q is less than or equal to the dimension of the state. Consider a $(q + 1)$-impulse solution. From Eq. (C.9):

$$\begin{aligned} \mathbf{y}(t_f) &= \mathbf{Z}(t_f, t_o)\mathbf{x}(t_o) + \sum_{k=1}^{q+1} \mathbf{Z}(t_f, t_k)\begin{bmatrix} \mathbf{0} \\ \Delta\mathbf{v}_k \end{bmatrix} \\ &= \mathbf{Z}(t_f, t_o)\mathbf{x}(t_o) + \sum_{k=1}^{q+1} \mathbf{R}(t_f, t_k)\Delta\mathbf{v}_k \end{aligned} \tag{C.13}$$

The matrix $\mathbf{Z}$ is formed by the q rows of the state transition matrix $\mathbf{\Phi}$ that correspond to the specified final state components and the matrix $\mathbf{R}$ is the partition comprising the right half of the matrix $\mathbf{Z}$.

The term on the right-hand side of Eq. (C.13) involving the $q+1$ velocity changes will now be investigated. For convenience, define

$$\mathbf{d} \equiv \mathbf{y}(t_f) - \mathbf{Z}(t_f, t_o)\mathbf{x}(t_o) = \sum_{k=1}^{q+1} \mathbf{R}(t_f, t_k)\Delta\mathbf{v}_k \tag{C.14}$$

where the vector $\mathbf{d}$ is determined by the specified initial and final conditions on the state. The velocity changes that satisfy Eq. (C.14) are to be determined. To that end define the q-dimensional vector

$$\mathbf{u}_k \equiv \mathbf{R}(t_f, t_k)\frac{\Delta\mathbf{v}_k}{\Delta v_k} \tag{C.15}$$

which represents the effect of a unit magnitude velocity change at time t_k. Combining Eqs. (C.14) and (C.15):

$$\mathbf{d} = \sum_{k=1}^{q+1} \Delta v_k \mathbf{u}_k \tag{C.16}$$

Because $\mathbf{u}_k$ is a q-dimensional vector the vectors $\mathbf{u}_1, \mathbf{u}_2, \cdots, \mathbf{u}_{q+1}$ are linearly dependent and there exists scalars $\alpha_1, \alpha_2, \cdots, \alpha_{q+1}$ (not all zero) such that

$$\sum_{k=1}^{q+1} \alpha_k \mathbf{u}_k = \mathbf{0} \tag{C.17}$$

Now let

$$\alpha \equiv \sum_{k=1}^{q+1} \alpha_k \tag{C.18}$$

where it can be assumed that $\alpha \geq 0$. (If $\alpha < 0$ change the signs of all the α_k.) Next, let

$$\beta_k \equiv \frac{\alpha_k}{\Delta v_k} \tag{C.19}$$

and choose an index r so that β_r is the maximum of all the β_k. Then $\beta_r \geq \beta_k$ for $k = 1, 2, \cdots, q+1$ and $\beta_r > 0$ because $\alpha \geq 0$. Multiplying Eq. (C.17) by $1/\beta_r$ and subtracting the result from Eq. (C.16) yields

$$\mathbf{d} = \sum_{k=1}^{q+1} \mu_k \mathbf{u}_k \tag{C.20}$$

where

$$\mu_k = \Delta v_k - \frac{\alpha_k}{\beta_r} = \frac{\Delta v_k}{\beta_r}(\beta_r - \beta_k) \tag{C.21}$$

Because $\beta_r \geq \beta_k$ it follows that $\mu_k \geq 0$ and also that $\mu_r = 0$. Finally, from Eqs. (C.15) and (C.20) it follows that

$$\mathbf{d} = \sum_{k=1}^{q+1} R(t_f, t_k)\Delta\tilde{\mathbf{v}}_k \tag{C.22}$$

where

$$\Delta\tilde{\mathbf{v}}_k = \frac{\mu_k}{\Delta v_k}\Delta\mathbf{v}_k \tag{C.23}$$

Equation (C.23) represents a q-impulse solution because $\mu_r = 0$ and therefore $\Delta\tilde{\mathbf{v}}_r = \mathbf{0}$. Note that the $\Delta\tilde{\mathbf{v}}_k$ are parallel to the original $\Delta\mathbf{v}_k$ and that the μ_k are the magnitudes of the $\Delta\tilde{\mathbf{v}}_k$.

The cost of the new q-impulse solution is

$$\tilde{J} = \sum_{k=1}^{q+1} \Delta\tilde{v}_k = \sum_{k=1}^{q+1} \mu_k = J - \frac{\alpha}{\beta_r}, \quad \text{where} \quad J = \sum_{k-1}^{q+1} \Delta v_k \tag{C.24}$$

Because $\alpha \geq 0$ and $\beta_r > 0$, it follows that $\tilde{J} \leq J$, indicating that the cost of the new q-impulse solution is equal to or less than the cost of the $(q+1)$-impulse solution.

Example C.1 Field-free space intercept problem

Consider the very simple system $\ddot{\mathbf{r}} = \mathbf{0}$. the solution is given by

$$\begin{aligned} \mathbf{r}(t) &= \mathbf{r}(\tau) + (t-\tau)\mathbf{v}(\tau) \\ \dot{\mathbf{r}}(t) &= \mathbf{v}(t) = \mathbf{v}(\tau) \end{aligned} \tag{C.25}$$

The state transition matrix for the state $\mathbf{x}^T = [\mathbf{r}^T\ \mathbf{v}^T]$ is then

$$\mathbf{\Phi}(t-\tau) = \begin{bmatrix} \mathbf{I} & (t-\tau)\mathbf{I} \\ \mathbf{0} & \mathbf{I} \end{bmatrix} \tag{C.26}$$

Consider a two-dimensional intercept problem. For this problem $q = 2$ because two final position components are specified. We will show that a 3-impulse solution can be reduced to a 2-impulse solution of lower cost. Let the impulse times be $t_o = 0, t_1 = 1$, $t_2 = 2$, and the final time be $t_f = 3$. Let the intercept condition (the target position at the final time) be $\mathbf{r}^T(3) = [10 \quad 7]$ and the vehicle initial conditions be $\mathbf{r}^T(0) = [1 \quad 1]$ and $\mathbf{v}^T(0) = [1 \quad 0]$.

In Eq. (C.13) $\mathbf{y}^T(3) = \mathbf{r}^T(3) = [10 \quad 7]$ and $\mathbf{x}^T(0) = [1 \quad 1 \quad 1 \quad 0]$. The transition matrix in Eq. (C.26) at the final time is

$$\boldsymbol{\Phi}(3) = \begin{bmatrix} \mathbf{I}_2 & 3\mathbf{I}_2 \\ \mathbf{0}_2 & \mathbf{I}_2 \end{bmatrix} \tag{C.27}$$

and in Eq. (C.13) $\mathbf{Z}(3) = [\mathbf{I}_2 \quad 3\mathbf{I}_2]$ from which $\mathbf{R}(3) = 3\mathbf{I}_2$. The reader can verify that in Eq. (C.14) $\mathbf{d} = \mathbf{y}(3) - \mathbf{Z}(3)\mathbf{x}(0)$ yields $\mathbf{d}^T = [10 \quad 7] - [4 \quad 1] = [6 \quad 6]$.

In Eq. (C.14) $\mathbf{R}(t_f, t_k) = \mathbf{R}(3 - t_k) = (3 - t_k)\mathbf{I}_2$. and $\mathbf{R}(3) = 3\mathbf{I}_2$, $\mathbf{R}(2) = 2\mathbf{I}_2$, $\mathbf{R}(1) = \mathbf{I}_2$. So

$$\mathbf{d} = 3\Delta\mathbf{v}_1 + 2\Delta\mathbf{v}_2 + \Delta\mathbf{v}_3 = \begin{bmatrix} 6 \\ 6 \end{bmatrix} \tag{C.28}$$

Three (nonunique) velocity changes that satisfy Eq. (C.28) are

$$\Delta\mathbf{v}_1 = \begin{bmatrix} 1 \\ 0 \end{bmatrix} \quad \Delta\mathbf{v}_2 = \begin{bmatrix} 0 \\ 1 \end{bmatrix} \quad \Delta\mathbf{v}_3 = \begin{bmatrix} 3 \\ 4 \end{bmatrix} \tag{C.29}$$

for which the cost $J = 7$.

Next, from Eq. (C.15) the $\mathbf{u}_k$ are calculated to be $\mathbf{u}_1^T = [3\ 0]$, $\mathbf{u}_2^T = [0\ 2]$, $\mathbf{u}_3^T = [3/5 \quad 4/5]$ and a (nonunique) set of α_k that satisfies Eq. (C.17) is $\alpha_1 = -1, \alpha_2 = -2$, $\alpha_3 = 5$, resulting in $\alpha = 2 > 0$ in Eq. (C.18).

Continuing with the procedure we calculate the β_k from Eq. (C.19) to be $\beta_1 = -1$, $\beta_2 = -2$, $\beta_3 = 1$ from which we determine that $\beta_r = \beta_3 = 1$. Next, the μ_k are determined from Eq. (C.21) to be $\mu_1 = 2$, $\mu_2 = 3$, $\mu_3 = 0$. with μ_3 equal to zero as it should be.

Finally, the velocity changes for the new 2-impulse solution are calculated from Eq. (C.23) as

$$\Delta\tilde{\mathbf{v}}_1 = 2\Delta\mathbf{v}_1 \qquad \Delta\tilde{\mathbf{v}}_2 = 3\Delta\mathbf{v}_2 \tag{C.30}$$

for which $\tilde{J} = 5$ in contrast to $J = 7$.

Thus a (nonoptimal) $q = 2$-impulse solution been determined that has a lower cost than the original 3-impulse solution. What is the optimal solution, you ask? For this example the optimal solution is a single impulse at $t = 0$ with $\Delta\mathbf{v}^T = [2 \quad 2]$ for a cost of $\sqrt{8} = 2.83$. And this provides an excellent example that while a solution to the problem requires no more than q impulses, the optimal solution may require fewer.

Figure C.1 displays the 2-impulse and 3-impulse solutions to this example problem. The solid line is the 3-impulse solution with impulses at $\mathbf{r}(0)$, $\mathbf{r}(1)$, and $\mathbf{r}(2)$ and the intercept at $\mathbf{r}(3)$. The dashed line is the 2-impulse solution of lower cost with the impulses at the same times $t = 0$ and $t = 1$ and a zero impulse at $t = 2$. The optimal single impulse solution is a straight line path from $\mathbf{r}(0)$ to $\mathbf{r}(3)$.

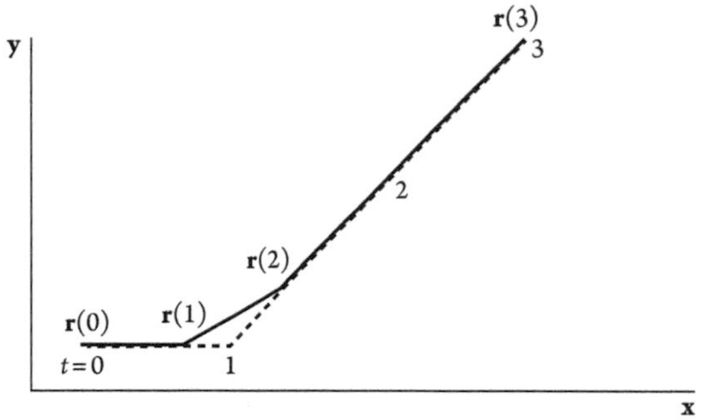

Figure C.1 Two- and Three-Impulse Field-Free Space Intercept Trajectories

In Ref. [C.1] another example is presented that is an undamped harmonic oscillator, which represents the out-of-plane motion near a circular reference orbit in the two-body problem. A rendezvous problem is analyzed, for which $q = 2$.

References

[C.1] Prussing, J.E., "Optimal Impulsive Linear Systems: Sufficient Conditions and Maximum Number of Impulses", *The Journal of the Astronautical Sciences*, Vol. 43, No. 2, Apr–Jun 1995, pp. 195–206.

[C.2] Edelbaum, T.N., "How Many Impulses?", *Astronautics and Aeronautics*, November 1967, pp. 64–9.

[C.3] Neustadt, L.W., "Optimization, a Moment Problem and Nonlinear Programming", *Journal of the Society for Industrial and Applied Mathematics, Series A Control*, Vol. 2, No. 1, 1964, pp. 33–53.

[C.4] Stern, R.G., and Potter, J.E., "Optimization of Midcourse Velocity Corrections", *Peaceful Uses of Automation in Outerspace*, Plenum Press, New York, 1966, pp. 70–83.

Appendix D
Linear System Theory

D.1 Homogeneous case

Consider the homogeneous linear system

$$\dot{\mathbf{y}}(t) = \mathbf{A}(t)\mathbf{y}(t) \tag{D.1}$$

where $\mathbf{y}(t)$ is an n-dimensional state vector. If $\mathbf{A}$ is a constant matrix the system is said to be autonomous or a constant-coefficient system, in contrast to a non-autonomous or time-varying system.

It is desired to determine the solution of Eq. (D.1), given initial conditions on the state vector at the initial time τ. Because the system is homogeneous, the solution is due only to initial conditions. And because the system is linear, we assume a solution of the form

$$\mathbf{y}(t) = \mathbf{\Phi}(t, \tau)\mathbf{y}(\tau) \tag{D.2}$$

Equation (D.2) is simply a statement that each component of $\mathbf{y}(t)$ is a linear combination of the components of the initial state $\mathbf{y}(\tau)$ with the coefficients being the elements of the matrix $\mathbf{\Phi}(t, \tau)$. This matrix is called the *state transition matrix* because it propagates the state at one time into the state at another time.

Before describing how this matrix is determined, we first describe some properties of the state transition matrix. There are six basic properties, as follows:

P1. Initial value:

$$\mathbf{\Phi}(\tau, \tau) = \mathbf{I} \tag{D.3}$$

where $\mathbf{I}$ is the $n \times n$ identity matrix. Property P1 simply states that if Eq. (D.2) is evaluated at $t = \tau$, then $\mathbf{y}(\tau)$ must be transformed into itself.

P2. Inversion property:

$$\mathbf{\Phi}(\tau, t) = \mathbf{\Phi}^{-1}(t, \tau) \tag{D.4}$$

Interchanging the arguments reverses the time direction of the transformation and is represented by the inverse of the state transition matrix. The mathematical justification for this property is that

$$\mathbf{y}(\tau) = \mathbf{\Phi}(\tau, t)\mathbf{y}(t) = \mathbf{\Phi}(\tau, t)\mathbf{\Phi}(t, \tau)\mathbf{y}(\tau) \tag{D.5}$$

The product of the two state transition matrices must be the identity matrix from Property P1, so one is the inverse of the other.

P3. Transition property:

$$\mathbf{\Phi}(t, \xi)\mathbf{\Phi}(\xi, \tau) = \mathbf{\Phi}(t, \tau) \tag{D.6}$$

This follows directly from

$$\mathbf{y}(t) = \mathbf{\Phi}(t, \xi)\mathbf{y}(\xi) = \mathbf{\Phi}(t, \xi)\mathbf{\Phi}(\xi, \tau)\mathbf{y}(\tau) \tag{D.7}$$

In other words, a transformation of the state from the initial time τ to the time ξ, followed by a transformation from time ξ to time t must be the same transformation as from the initial time τ to time t.

P4. Determinant property:

$$\det \mathbf{\Phi}(t, \tau) = \exp\left\{\int_{\tau}^{t} \mathrm{tr}[\mathbf{A}(\xi)]d\xi\right\} \neq 0 \tag{D.8}$$

where the symbol "tr" represents the trace of the matrix $\mathbf{A}(t)$ (equal to the sum of its diagonal elements). Note that this property guarantees that $\mathbf{\Phi}(t, \tau)$ is nonsingular because the determinant is positive. A proof of this property will not be given here, but the property follows from the fact that the determinant satisfies the equation (See Ref. [D.1]):

$$\frac{d}{dt} \det \mathbf{\Phi}(t, \tau) = \mathrm{tr}[\mathbf{A}(t)] \det \mathbf{\Phi}(t, \tau) \tag{D.9a}$$

with initial condition provided by Eq. (D.3):

$$\det \mathbf{\Phi}(\tau, \tau) = \det \mathbf{I} = 1 \tag{D.9b}$$

This property can be very useful as a check on the accuracy of a numerical solution for $\mathbf{\Phi}(t, \tau)$. For example, if $\mathrm{tr}[\mathbf{A}] = 0$, the determinant is identically 1. The determinant also has a geometrical interpretation related to a volume element in the n-dimensional state space. Consider a collection of initial states occupying a *unit* volume. The value of the determinant is the volume occupied by those states at a later time. This volume will decrease if the trajectories asymptotically converge due to damping, stay equal to unity, or it can increase due to instability. From Eq. (D.9a) the determinant increases if $tr[\mathbf{A}] > 0$, remains constant if $tr[\mathbf{A}] = 0$, and decreases if $tr[\mathbf{A}] < 0$.

P5. Time invariant property:
If the matrix **A** is constant, then

$$\mathbf{\Phi}(t,\tau) = \mathbf{\Phi}(t-\tau) \tag{D.10}$$

This is a result of the fact that if the system characteristics are time-invariant, the response at time t due to initial conditions at time τ depends only on the *elapsed time* $t - \tau$. This property is useful in solving for $\mathbf{\Phi}$ because one can assume $\tau = 0$, obtain the solution $\mathbf{\Phi}(t)$, and then replace t everywhere by $t - \tau$ to obtain $\mathbf{\Phi}(t-\tau)$.

P6. Transition matrix equation:

$$\frac{d}{dt}\mathbf{\Phi} = \dot{\mathbf{\Phi}}(t,\tau) = \mathbf{A}(t)\mathbf{\Phi}(t,\tau) \tag{D.11}$$

To derive this result one first calculates (assuming τ to be constant):

$$\dot{\mathbf{y}}(t) = \dot{\mathbf{\Phi}}(t,\tau)\mathbf{y}(\tau) \tag{D.12}$$

where $\mathbf{y}(\tau)$ is considered to be a constant initial state vector. Substituting Eq. (D.12) into Eq. (D.1) yields

$$\dot{\mathbf{\Phi}}(t,\tau)\mathbf{y}(\tau) = \mathbf{A}(t)\mathbf{y}(t) \tag{D.13}$$

and, using Eq. (D.2)

$$[\dot{\mathbf{\Phi}}(t,\tau) - \mathbf{A}(t)\mathbf{\Phi}(t,\tau)]\mathbf{y}(\tau) = \mathbf{0} \tag{D.14a}$$

Because $\mathbf{y}(\tau)$ is arbitrary it is necessary that Eq. (D.11) be satisfied in order for Eq. (D.14a) to be satisfied for all initial states $\mathbf{y}(\tau)$. Equation (D.11) can be solved using the initial condition (D.3) to determine the state transition matrix $\mathbf{\Phi}(t,\tau)$.

There are various ways of determining the state transition matrix, depending on the situation. In the most general time-varying case one simply numerically integrates Eq. (D.11) using (D.3) as an initial condition and (D.9a) as a numerical accuracy check. To numerically integrate, one treats each column of $\mathbf{\Phi}(t,\tau)$ as an n –vector $\boldsymbol{\phi}^{(k)}$, $k = 1, 2, \cdots, n$:

$$\dot{\boldsymbol{\phi}}^{(k)}(t,\tau) = \mathbf{A}(t)\boldsymbol{\phi}^{(k)}(t,\tau) \tag{D.14b}$$

with initial condition equal to the k-th column of the identity matrix, i.e.,

$$\Phi_{jk}(\tau,\tau) = \phi_j^{(k)}(\tau,\tau) = \delta_{jk} \tag{D.14c}$$

To simultaneously determine all the columns, one concatenates these n vectors into a single $n^2 \times 1$ vector equation to be integrated once.

If the matrix **A** is constant the solution to Eq. (D.11) can be formally represented as

$$\boldsymbol{\Phi}(t) = e^{At} \tag{D.15}$$

where the right-hand side is an $n \times n$ matrix which requires some effort to evaluate directly.

A simpler method for a constant matrix **A** involves using Laplace transforms. Transforming Eq. (D.11)

$$s\mathbf{Y}(s) - \mathbf{y}(0^-) = \mathbf{A}\,\mathbf{Y}(s) \tag{D.16}$$

where $\mathbf{Y}(s)$ is the Laplace transform of $\mathbf{y}(t)$ and $\mathbf{y}(0^-)$ is the initial state at $t = 0^-$ (just prior to $t = 0$). Solving Eq. (D.16) for $\mathbf{Y}(s)$:

$$\mathbf{Y}(s) = (s\mathbf{I} - \mathbf{A})^{-1}\mathbf{y}(0^-) \tag{D.17}$$

where the Laplace variable s has been multiplied by the identity matrix in order to be able to subtract the matrix **A** from it. By comparison with Eq. (D.2) with $\tau = 0$ and using the property in (D.10), it is evident that the matrix $(s\mathbf{I} - \mathbf{A})^{-1}$ is the Laplace transform of the state transition matrix $\boldsymbol{\Phi}(t)$. This, along with the property described after Eq. (D.10), represents a method for determining $\boldsymbol{\Phi}(t)$ for a given constant matrix **A**; the inverse Laplace transform of $(s\mathbf{I} - \mathbf{A})^{-1}$ is the state transition matrix $\boldsymbol{\Phi}(t)$. The matrix $\Phi(t - \tau)$ is obtained from $\Phi(t)$ by simply replacing t by $t - \tau$.

Example D.1 Double integrator state transition matrix

a) Determine the coefficient matrix **A**.

For $\ddot{y} = 0$ define

$$\mathbf{y} = \begin{bmatrix} y_1 \\ y_2 \end{bmatrix}$$

Then, in first order form $\dot{y}_1 = y_2$ and $\dot{y}_2 = 0$.

So $\dot{\mathbf{y}} = \mathbf{A}\mathbf{y}$ with

$$\mathbf{A} = \begin{bmatrix} 0 & 1 \\ 0 & 0 \end{bmatrix}$$

b) Determine the state transition matrix.

$$s\mathbf{I} - \mathbf{A} = \begin{bmatrix} s & -1 \\ 0 & s \end{bmatrix}$$

Then

$$(s\mathbf{I} - \mathbf{A})^{-1} = \frac{1}{s^2}\begin{bmatrix} s & 1 \\ 0 & s \end{bmatrix} = \begin{bmatrix} 1/s & 1/s^2 \\ 0 & 1/s \end{bmatrix}$$

So the inverse transform yields

$$\Phi(t) = \begin{bmatrix} 1 & t \\ 0 & 1 \end{bmatrix}$$

and

$$\Phi(t-\tau) = \begin{bmatrix} 1 & t-\tau \\ 0 & 1 \end{bmatrix}$$

D.2 Inhomogeneous case

Consider the inhomogeneous linear system

$$\dot{\mathbf{y}}(t) = \mathbf{A}(t)\mathbf{y}(t) + \mathbf{B}(t)\mathbf{u}(t) \tag{D.18}$$

where the inhomogeneous term $\mathbf{B}(t)\mathbf{u}(t)$ occurs due to control of the system. Consider the vector $\mathbf{u}(t)$ to be $m \times 1$, meaning that $\mathbf{B}(t)$ is an $n \times m$ matrix.

The response of the inhomogeneous system is due to both initial conditions and the control input $\mathbf{u}(t)$. To determine the solution to (D.18), generalize (D.2) using the Variation of Parameters Method:

$$\mathbf{y}(t) = \mathbf{\Phi}(t,\tau)\mathbf{c}(t) \tag{D.19}$$

where the constant initial condition $\mathbf{y}(\tau)$ has been replaced by an unknown time-varying vector $\mathbf{c}(t)$ to be determined. Differentiating (D.19)

$$\begin{aligned} \dot{\mathbf{y}}(t) &= \dot{\mathbf{\Phi}}(t,\tau)\mathbf{c}(t) + \mathbf{\Phi}(t,\tau)\dot{\mathbf{c}}(t) \\ &= \mathbf{A}(t)\mathbf{\Phi}(t,\tau)\mathbf{c}(t) + \mathbf{\Phi}(t,\tau)\dot{\mathbf{c}}(t) \end{aligned} \tag{D.20}$$

where (D.11) has been used.

Comparing (D.20) with (D.18–19) it must be that

$$\mathbf{\Phi}(t,\tau)\dot{\mathbf{c}}(t) = \mathbf{B}(t)\mathbf{u}(t) \tag{D.21}$$

and one can solve for $\mathbf{c}(t)$ using (D.4):

$$\dot{\mathbf{c}}(t) = \mathbf{\Phi}(\tau,t)\mathbf{B}(t)\mathbf{u}(t) \tag{D.22}$$

Equation (D.22) can be integrated to yield

$$\mathbf{c}(t) = \mathbf{c}(\tau) + \int_{\tau}^{t} \boldsymbol{\Phi}(\tau,\xi)\mathbf{B}(\xi)\mathbf{u}(\xi)d\xi \tag{D.23}$$

But, from (D.2) and (D.19)

$$\mathbf{c}(\tau) = \mathbf{y}(\tau) \tag{D.24}$$

so that, combining (D.23–24) with (D.19):

$$\mathbf{y}(t) = \boldsymbol{\Phi}(t,\tau)\mathbf{y}(\tau) + \int_{\tau}^{t} \boldsymbol{\Phi}(t,\xi)\mathbf{B}(\xi)\mathbf{u}(\xi)d\xi \tag{D.25}$$

where (D.6) has been used.

The second term in Eq. (D.25) is a *convolution integral* that represents the particular solution to (D.18) which is added to the homogeneous solution (D.2) to yield the general solution in the inhomogeneous case.

Using Laplace transforms for constant **A** and **B**, Eq. (D.17) becomes

$$\mathbf{Y}(s) = (s\mathbf{I} - \mathbf{A})^{-1}\mathbf{y}(0^{-}) + \mathbf{H}(s)\mathbf{U}(s) \tag{D.26}$$

where $\mathbf{U}(s)$ is the Laplace transform of $\mathbf{u}(t)$ and $\mathbf{H}(s) = (s\mathbf{I} - \mathbf{A})^{-1}\mathbf{B}$ is the *transfer function* of the system.

Problems

D.1 For a constant matrix **A** show that

a) $\Phi(t_1 + t_2) = \Phi(t_1)\Phi(t_2)$

b) $\Phi^{-1}(t) = \Phi(-t)$

D.2 Given $\Phi(t,\tau)$, determine $\mathbf{A}(t)$.

D.3 a) Determine the coefficient matrix **A** for the system $\ddot{y} + y = 0$.

b) Using Laplace transforms determine the corresponding state transition matrix $\Phi(t-\tau)$.

Reference

[D.1] Wiberg, D.M., *State Space and Linear Systems*, Schaum's Outline Series, McGraw-Hill, Section 5.1, 1971

Appendix E
Maximum Range Using Continuous Thrust in a Uniform Gravitational Field

As shown in Fig. E.1 the vehicle starts at rest at $t = 0$ at the origin of an $x - y$ coordinate frame with a constant gravitational acceleration acting in the negative y direction. This is a "flat earth" gravitational model that is a reasonable approximation if the changes in x and y are fairly small.

The first phase is a thrust arc ending at the burnout time t_b when all the propellant has been consumed. This is followed by a no-thrust (free-fall or ballistic) that impacts $y = 0$ at time T and location $x(T)$, which is the range to be maximized by using optimal thrust magnitude and direction.

The analysis here is based on Section 2.2 of Ref. [E.1], with more details and some enhancements.

The equations of motion are

$$\dot{\mathbf{r}} = \mathbf{v} \tag{E.1a}$$

$$\dot{\mathbf{v}} = \mathbf{g}_o + \Gamma \mathbf{u} \tag{E.1b}$$

In Eqs. (1) $\mathbf{r}^T = [x\ y]$ and $\mathbf{v}^T = [\mathrm{v}_x\ \mathrm{v}_y]$ and the initial conditions are $\mathbf{r}(0) = \mathbf{v}(0) = \mathbf{0}$. The thrust acceleration is represented by its magnitude Γ and a unit vector in the thrust direction $\mathbf{u}$.

For maximum range the cost to be minimized is the Mayer form $J = -x(T)$ and the final time T is defined by $y(T) = 0$.

Define the Hamiltonian as

$$H = \boldsymbol{\lambda}_r^T \mathbf{v} + \boldsymbol{\lambda}_v^T (\mathbf{g}_o + \Gamma \mathbf{u}) \tag{E.2}$$

As discussed in Section 4.1 the Hamiltonian is minimized over the choice of thrust direction by aligning $\mathbf{u}$ opposite to $\boldsymbol{\lambda}_v$. Defining the primer vector $\mathbf{p}(t) \equiv -\boldsymbol{\lambda}_v(t)$ we have the optimal thrust direction $\mathbf{u}(t) = \mathbf{p}(t)/p(t)$. The primer vector satisfies a special case of Eq. (4.15):

$$\ddot{\mathbf{p}} = \mathbf{G}\mathbf{p} = \mathbf{0} \tag{E.3}$$

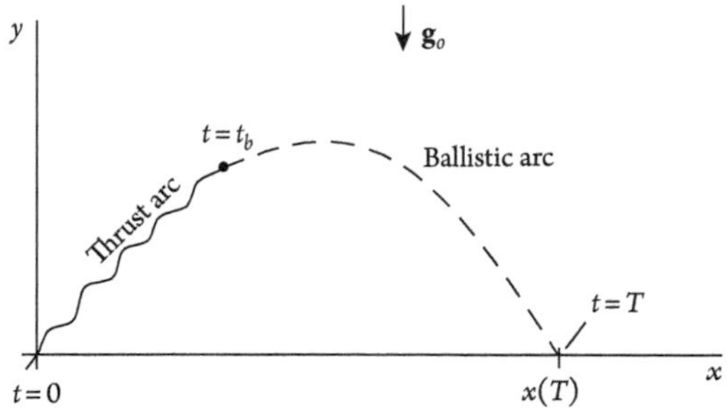

Figure E.1 Continuous thrust trajectory

because the gravity gradient matrix is a zero matrix for a uniform gravitational field. The solution to Eq. (E.3) is

$$\mathbf{p}(t) = \mathbf{p}_o + t\dot{\mathbf{p}}_o; \dot{\mathbf{p}}(t) = \dot{\mathbf{p}}_o \tag{E.4}$$

The conditions at time t_b are the "terminal conditions" for the thrust arc and there are no terminal constraints, but only the terminal cost $\phi = -x(T)$. The boundary conditions on the primer vector and its derivative are modified versions of Eqs. (4.16) and (4.17):

$$\mathbf{p}^T(t_b) = -\frac{\partial \phi}{\partial \mathbf{v}(t_b)} \tag{E.5a}$$

and

$$\dot{\mathbf{p}}^T(t_b) = \frac{\partial \phi}{\partial \mathbf{r}(t_b)} \tag{E.5b}$$

Using the optimal thrust direction and the primer vector definition the Hamiltonian of Eq. (4.15) becomes

$$H = \dot{\mathbf{p}}^T\mathbf{v} - \mathbf{p}^T\mathbf{g}_o - p\Gamma \tag{E.6}$$

With a bounded thrust acceleration $0 \leq \Gamma \leq \Gamma_m$ the Hamiltonian in Eq. (E.6) is minimized by choosing $\Gamma = \Gamma_m$ because the primer magnitude $p > 0$. (We will assume a constant Γ_m but note that even if it varies with time we will always choose the maximum value.) The product $\Gamma_m t_b$ is equal to the available fuel and is the "Δv" for the maneuver.

Returning to the equations of motion (E.1a) and (E.1b), v_x is constant for $t > t_b$ because there is no gravitational force in the x-direction and

$$x(T) = x(t_b) + v_x(t_b)(T - t_b) \tag{E.7}$$

and we need to solve for $T - t_b$.

For $t > t_b$ we have

$$\ddot{y} = -g_o \tag{E.8a}$$

and integrating,

$$\dot{y} = -g_o(t - t_b) + v_y(t_b) \tag{E.8b}$$

$$y = -\frac{1}{2}g_o(t - t_b)^2 + (t - t_b)v_y(t_b) + y(t_b) \tag{E.8c}$$

We next impose the condition $y(T) = 0$ to solve for $T - t_b$:

$$(T - t_b)^2 - \frac{2v_y(t_b)}{g_o}(T - t_b) - \frac{2y(t_b)}{g_o} = 0 \tag{E.9}$$

Applying the quadratic formula yields

$$T - t_b = \frac{v_y(t_b) \pm s}{g_o} \tag{E.10}$$

where $s \equiv [v_y^2(t_b) + 2g_o y(t_b)]^{\frac{1}{2}}$. Because $s > |v_y(t_b)|$ choosing the "+" sign will yield $T - t_b > 0$.

Substituting into Eq. (E.7) we have an expression for our cost:

$$J = \phi = -x(t_b) - \frac{v_x(t_b)}{g_o}[v_y(t_b) + s] \tag{E.11}$$

We can now evaluate the boundary conditions of Eqs. (5). First, $\partial s/\partial y(t_b) = g_o/s$. Then,

$$\dot{\mathbf{p}}^T(t_b) = [-1 \quad -v_x(t_b)/s] \tag{E.12a}$$

and

$$\mathbf{p}^T(t_b) = \left(\frac{1}{g_o}[v_y(t_b) + s] \quad \frac{v_x(t_b)}{g_o s}[v_y(t_b) + s]\right) \tag{E.12b}$$

Note that

$$\mathbf{p}(t_b) = -\frac{1}{g_o}[v_y(t_b) + s]\dot{\mathbf{p}}(t_b) \tag{E.13}$$

and since $s > |v_y(t_b)|$ [see after Eq. (E.10)], $\mathbf{p}(t_b)$ is in the opposite direction of $\dot{\mathbf{p}}(t_b)$. Based on Eq. (E.4) this means that for $0 \leq t \leq t_b$, $\mathbf{p}(t)$ is directed opposite to the constant vector $\dot{\mathbf{p}}(t_b)$, i.e., the thrust direction is *constant*.

The value of the constant thrust angle β_o is given by

$$\tan\beta_o = \frac{p_y(t_b)}{p_x(t_b)} = \frac{v_x(t_b)}{s} \tag{E.14}$$

Based on $\dot{v}_y = \Gamma_m \sin\beta_o - g_o$, we determine that $v_y(t_b) = t_b(\Gamma_m \sin\beta_o - g_o)$ and $y(t_b) = t_b^2(\Gamma \sin\beta_o - g_o)/2$. Using these expressions, $v_x(t_b) = t_b\Gamma_m \cos\beta_o$, and the definition of s after Eq. (E.10), Eq. (E.14) becomes

$$\tan\beta_o = \frac{t_b\Gamma_m \cos\beta_o}{[t_b^2(\Gamma_m \sin\beta_o - g_o]^2 + g_o t_b^2(\Gamma_m \sin\beta_o - g_o)]^{\frac{1}{2}}} \tag{E.15}$$

After some algebra we arrive at

$$\sin^3\beta_o + \frac{\Gamma_m}{g_o}\cos 2\beta_o = 0 \tag{E.16}$$

Similarly, we can determine an expression for the flight path angle, γ_o:

$$\tan\gamma_o = \frac{y(t_b)}{x(t_b)} = \tan\beta_o - \frac{g_o}{\Gamma_m}\sec\beta_o \tag{E.17}$$

For values of $\Gamma_m/g_o > 1$ (for which the vehicle can lift off the launch pad) there are two solutions to Eq. (E.16) that are symmetric about $\pi/2$, corresponding to motion to the right or left in Fig. E.1.

Table E.1 shows, for various values of Γ_m/g_o, the thrust angle, flight path angle, and the range nondimensionalized by the ballistic projectile range $v_o{}^2/g_o = (\Gamma_m t_b)^2/g_o$. Note

Table E.1 Maximum range for continuous thrust in a uniform gravity field.

thrust acceleration Γ_m/g_o	thrust angle β_o(degrees)	path angle γ_o (degrees)	nondimensional range $g_o x(T)/\Gamma_m^2 t_b^2$
1.000	90.	0.	0.
1.001	87.4	0.00526	0.0224
1.1	68.6	3.45	0.226
1.5	56.3	16.6	0.482
2.0	52.1	25.2	0.625
3.0	49.2	32.9	0.756
10.0	46.1	41.8	0.929
20.0	45.5	43.4	0.965
–	–	–	–
∞	45.0	45.0	1.0

that the values of the range are less than one due to the gravity loss incurred by the continuous thrust. Also, as $\Gamma_m/g_o \to \infty$ the thrust angle and the flight path angle become equal to the well-known maximum range solution for the ballistic problem. In the limit as $\Gamma_m \to \infty$, $t_b \to 0$ and the product approaches the launch velocity v_o.

Reference

[E.1] Lawden, D.F., *Optimal Trajectories for Space Navigation*, Butterworths, London, 1963.

Appendix F
Quadratic Forms

F.1 Matrix symmetry

Consider the (scalar) quadratic form $\mathbf{x}^T M\mathbf{x}$ where $\mathbf{x}$ is an $n \times 1$ vector and M is an $n \times n$ matrix. The matrix M can be assumed to be symmetric (or replaced by its symmetric part). To show this let

$$M = M_S + M_A \tag{F.1}$$

where

$$M_S \equiv \frac{1}{2}(M + M^T) \tag{F.2}$$

is symmetric because $M_S^T = M_S$ and

$$M_A \equiv \frac{1}{2}(M - M^T) \tag{F.3}$$

is anti- (skew-) symmetric because $(M_A^T = -M_A)$.

Using Eq. (F.1)

$$\mathbf{x}^T M\mathbf{x} = \mathbf{x}^T M_S\mathbf{x} + \mathbf{x}^T M_A\mathbf{x} \tag{F.4}$$

Consider the last term in Eq. (F.4) and note that

$$(\mathbf{x}^T M_A\mathbf{x})^T = \mathbf{x}^T M_A^T\mathbf{x} = -\mathbf{x}^T M_A\mathbf{x} = \mathbf{x}^T M_A\mathbf{x} \tag{F.5}$$

because M_A is anti-symmetric and the transpose of a scalar is equal to the scalar. From Eq. (F.5)

$$2\mathbf{x}^T M_A\mathbf{x} = 0 \Longrightarrow \mathbf{x}^T M_A\mathbf{x} = 0 \tag{F.6}$$

Therefore, from Eq. (F.4) we have

$$\mathbf{x}^T M \mathbf{x} = \mathbf{x}^T M_S \mathbf{x} \tag{F.7}$$

showing that M can always be assumed to be symmetric (or replaced by its symmetric part M_S).

F.2 The derivative of a quadratic form

To determine

$$\frac{\partial}{\partial \mathbf{x}} \left(\frac{1}{2} \mathbf{x}^T M \mathbf{x} \right) \tag{F.8}$$

let $\mathbf{y} \equiv M\mathbf{x}$. Then the derivative in Eq. (F.8) becomes

$$\frac{\partial}{\partial \mathbf{x}} \left(\frac{1}{2} \mathbf{x}^T \mathbf{y} \right) = \frac{1}{2} \left(\mathbf{x}^T \frac{\partial \mathbf{y}}{\partial \mathbf{x}} + \mathbf{y}^T \frac{\partial \mathbf{x}}{\partial \mathbf{x}} \right) \tag{F.9}$$

which is equal to

$$\frac{1}{2} (\mathbf{x}^T M + \mathbf{x}^T M I) = \mathbf{x}^T M \tag{F.10}$$

where the symmetry of the matrix M has been used. So the result is that

$$\frac{\partial}{\partial \mathbf{x}} \left(\frac{1}{2} \mathbf{x}^T M \mathbf{x} \right) = \mathbf{x}^T M \tag{F.11}$$

Appendix G
Simple Conjugate Point Example

A classical geometrical example of a conjugate point is the minimum distance path between two points on a sphere. A great circle connecting the two points satisfies the first-order NC, but a conjugate point exists at a point diametrically opposite the final point. That problem is treated in Example 2 of Ref. [8.2] and is somewhat complicated. The example treated here is much simpler, but demonstrates the essential characteristics of a conjugate point and a new procedure. What follows is a summary of Ref. [G.1].

A simple example of a conjugate point is the two-dimensional problem of the minimum time path of a vehicle moving at constant speed from a point to a circle in the $x_1 - x_2$ plane. (This is equivalent to a minimum distance problem, but sounds more exciting.)

For a time interval $0 \leq t \leq t_f$ the cost is defined to be $J = \phi\left[\mathbf{x}(t_f), t_f\right] = t_f$ and the initial point is defined in terms of the two state variables as a point on the x_1 axis:

$$\mathbf{x}(0) = \begin{bmatrix} x_1(0) \\ x_2(0) \end{bmatrix} = \begin{bmatrix} a \\ 0 \end{bmatrix} \tag{G.1}$$

The equation of motion (for a unit vehicle speed) is

$$\dot{\mathbf{x}} = \mathbf{f}(\mathbf{x}, \mathbf{u}, t) = \begin{bmatrix} \cos\theta \\ \sin\theta \end{bmatrix} \tag{G.2}$$

where θ is the heading angle (the control variable). The terminal constraint is the unit circle:

$$\psi\left[\mathbf{x}(t_f), t_f\right] = \frac{1}{2}\left[x_1^2(t_f) + x_2^2(t_f) - 1\right] = 0 \tag{G.3}$$

Consider the case in which $0 < a < 1$ in Eq. (G.1), i.e., the initial point is inside the circle, but not at the origin. In the notation of Chapter 8, $n = 2$ state variables x_1 and x_2, $m = 1$ control variable θ and $q = 0$ represents the single terminal constraint.

To apply the first-order NC the Hamiltonian function of Eq. (3.7) is defined as

$$H(\mathbf{x}, \mathbf{u}, \boldsymbol{\lambda}, t) = L + \boldsymbol{\lambda}^T \mathbf{f} = \lambda_1 \cos\theta + \lambda_2 \sin\theta \tag{G.4}$$

where $L \equiv 0$ in this example. An augmented terminal function of Eq. (3.36) is defined as

$$\Phi[\mathbf{x}(t_f), t_f, \nu] \equiv \phi + \nu\psi = t_f + \frac{\nu}{2}[x_1^2(t_f) + x_2^2(t_f) - 1] \tag{G.5}$$

Applying the first-order NC of Eqs. (3.12–3.14):

$$\frac{\partial H}{\partial \theta} = 0 = -\lambda_1 \sin\theta + \lambda_2 \cos\theta \tag{G.6}$$

which yields

$$\tan\theta = \frac{\lambda_2}{\lambda_1} \tag{G.7}$$

Next,

$$\dot{\boldsymbol{\lambda}}^T = -\frac{\partial H}{\partial \mathbf{x}} = \mathbf{0}^T \tag{G.8}$$

indicating a constant $\boldsymbol{\lambda}(t)$ which implies a constant heading angle θ. The boundary condition for Eq. (G.8) is

$$\boldsymbol{\lambda}^T(t_f) = \frac{\partial \boldsymbol{\Phi}}{\partial \mathbf{x}(t_f)} = \nu[x_1(t_f) \quad x_2(t_f)] \tag{G.9}$$

The solution to Eq. (G.2) for constant θ evaluated at the final time is

$$x_1(t_f) = a + t_f \cos\theta \tag{G.10a}$$

$$x_2(t_f) = t_f \sin\theta \tag{G.10b}$$

And for a constant θ Eqs. (G.7–G.10b) yield

$$\tan\theta = \frac{x_2(t_f)}{x_1(t_f)} = \frac{t_f \sin\theta}{a + t_f \cos\theta} \tag{G.11}$$

which implies that $\sin\theta = 0$ for $a \neq 0$. And thus the values of θ that satisfy the NC are 0 and π, for which $x_2(t_f) = 0$ and $x_1(t_f) = 1$ or -1, respectively.

Because the final time t_f is unspecified there is an additional NC of Eq. (3.46) $\Omega = d\Phi/dt_f + L(t_f) = 0$ which can be calculated using Eq. (3.47) as $\partial\Phi/\partial t_f + H(t_f) = 0 = 1 + \nu x_1(t_f)\cos\theta$, which yields $\nu = -1$ for both $\theta = 0$ and π.

To apply the second-order conditions, from Eq. (G.4) we have from Eq. (G.6)

$$H_{\theta\theta} = -\lambda_1 \cos\theta - \lambda_2 \sin\theta \tag{G.12}$$

and from Eq. (G.9) for $\theta = 0, \lambda_1 = \nu x_1(t_f) = -1$ and for $\theta = \pi, \lambda_1 = \nu x_1(t_f) = 1$, so $H_{\theta\theta}$ is equal to 1 for both cases. Both the Legendre-Clebsch condition and the Strengthened Legendre-Clebsch conditions are satisfied.

Next we test for the existence of a conjugate point. Part of the procedure for determining a conjugate point is the terminal constraint nontangency condition of Eq. (8.37) $d\psi/dt_f \neq 0$. Because ψ in Eq. (G.3) is not an explicit function of t_f the derivative is calculated as

$$\frac{d\psi}{dt_f} = \frac{\partial\psi}{\partial x_1(t_f)}\dot{x}_1(t_f) + \frac{\partial\psi}{\partial x_2(t_f)}\dot{x}_2(t_f) = x_1(t_f)\cos\theta + x_2(t_f)\sin\theta = 1 \tag{G.13}$$

for both $\theta = 0$ and $\theta = \pi$, so the condition $d\psi/dt_f \neq 0$ is satisfied.

Next we need to determine a 4×4 transition matrix $\Theta(t, t_f)$ that satisfies Eq. (8.14b–c):

$$\dot{\Theta}(t, t_f) = P(t)\Theta(t, t_f); \quad \Theta(t_f, t_f) = I_4 \tag{G.14a}$$

where

$$P = \begin{bmatrix} A_1 & -A_2 \\ -A_o & -A_1^T \end{bmatrix} \tag{G.14b}$$

and the 2×2 partitions in Eq. (G.14b) are given by Eqs. (8.11b, 8.11c, and 8.12c):

$$A_o = H_{xx} - H_{x\theta}H_{\theta\theta}^{-1}H_{\theta x} \tag{G.15a}$$

$$A_1 = \mathbf{f}_x - \mathbf{f}_\theta H_{\theta\theta}^{-1}H_{\theta x} \tag{G.15b}$$

$$A_2 = \mathbf{f}_\theta H_{\theta\theta}^{-1}\mathbf{f}_\theta^T \tag{G.15c}$$

In this example problem $\mathbf{f}_x = 0_2$ (a 2×2 zero matrix), and $H_{\theta x} = 0_2$, so $A_o = A_1 = 0_2$. The matrix A_2 is (using $\sin\theta = 0$)

$$A_2 = \begin{bmatrix} 0 & 0 \\ 0 & 1 \end{bmatrix} \tag{G.16}$$

for both $\theta = 0$ and π. So in Eq. (G.14b)

$$P = \begin{bmatrix} 0 & 0 & 0 & 0 \\ 0 & 0 & 0 & -1 \\ 0 & 0 & 0 & 0 \\ 0 & 0 & 0 & 0 \end{bmatrix} \tag{G.17}$$

and the corresponding transition matrix in Eq. (G.14a) is

$$\Theta(t, t_f) = \begin{bmatrix} 1 & 0 & 0 & 0 \\ 0 & 1 & 0 & t_f - t \\ 0 & 0 & 1 & 0 \\ 0 & 0 & 0 & 1 \end{bmatrix} \tag{G.18}$$

which defines 2×2 partitions

$$\Theta(t, t_f) = \begin{bmatrix} \Theta_{11} & \Theta_{12} \\ \Theta_{21} & \Theta_{22} \end{bmatrix} \tag{G.19}$$

Finally, a conjugate point exists if the 2×2 matrix $X(t)$ is singular, where

$$X(t) = \Theta_{11} + \Theta_{12} S_F \tag{G.20}$$

The matrix S_F is given by Eq. (8.38), which in this simple example reduces to $S_F = \nu I_2 = -I_2$. Evaluating Eq. (G.20) using Eq. (G.18)

$$X(t) = \begin{bmatrix} 1 & 0 \\ 0 & 1 \end{bmatrix} + \begin{bmatrix} 0 & 0 \\ 0 & t_f - t \end{bmatrix} \begin{bmatrix} -1 & 0 \\ 0 & -1 \end{bmatrix} = \begin{bmatrix} 1 & 0 \\ 0 & 1 + t - t_f \end{bmatrix} \tag{G.21}$$

and

$$\det X(t) = 1 + t - t_f \tag{G.22}$$

At this point the two cases $\theta = 0$ and $\theta = \pi$ must be considered separately.

For $\theta = 0$:

$x_1(t_f) = 1$ and from Eq. (G.10a) $t_f = 1 - a$. From Eq. (G.22) $\det X(t) = 0$ for $t = -a$, but that is a negative value and is outside the time interval $[0, 1 - a]$ for this solution. Therefore, no conjugate point exists for $\theta = 0$.

For $\theta = \pi$:

$x_1(t_f) = -1$ and from Eq. (G.10a) $t_f = 1 + a$. From Eq. (G.22) $\det X(t) = 0$ for $t = a$, which is the time that the vehicle crosses the origin. Therefore a conjugate point exists at the origin for $\theta = \pi$, the solution is not the minimum time solution, and there exists a neighboring solution that requires less time.

Such a lower cost neighboring solution is obtained by simply bypassing the origin and avoiding the conjugate point. Consider a straight line path from the initial point to the terminal constraint with a final time $\hat{t}_f$. Let $x_2(\hat{t}_f)$ be a small positive number ε rather than 0. Then from Eq. (G.3) $x_1^2(\hat{t}_f) = 1 - \varepsilon^2$. So $x_1(\hat{t}_f) = (1 - \varepsilon^2)^{\frac{1}{2}}$ which is approximately $1 - \frac{1}{2}\varepsilon^2$. The time $\hat{t}_f$ to get to this point (which is equal to the distance traveled) is given by

$$\hat{t}_f^2 = \varepsilon^2 + \left(1 + a - \frac{1}{2}\varepsilon^2\right)^2 \tag{G.23a}$$

which is approximately

$$\hat{t}_f \approx \varepsilon^2 + (1 + a)^2 - \varepsilon^2(1 + a) = (1 + a)^2 - \varepsilon^2 a < (1 + a)^{2.} \tag{G.23b}$$

Therefore $\hat{t}_f < 1 + a$, and is a smaller time than $t_f = 1 + a$ on the stationary solution containing the conjugate point.

Note also a typical property of a conjugate point. There are infinitely many paths from the conjugate point (the origin) to the terminal constraint (the circle centered at the origin) all having the same cost, but they are not part of the optimal solution if $0 < a < 1$.

Reference

[G.1] Prussing, J.E., "Simplified Conjugate Point Procedure", *Journal of Guidance, Control, and Dynamics*, Vol. 40, No. 5, 2017, pp. 1255–57.

Index

Locators in *italics* refer to figures